EXPLORING SCIENCE IN 36 WEEKS

4 in 1 Workbook

DAILY SCIENCE WORKBOOK

for Kids in **Grades 4-5** To Master the Subject and Ace the Tests!

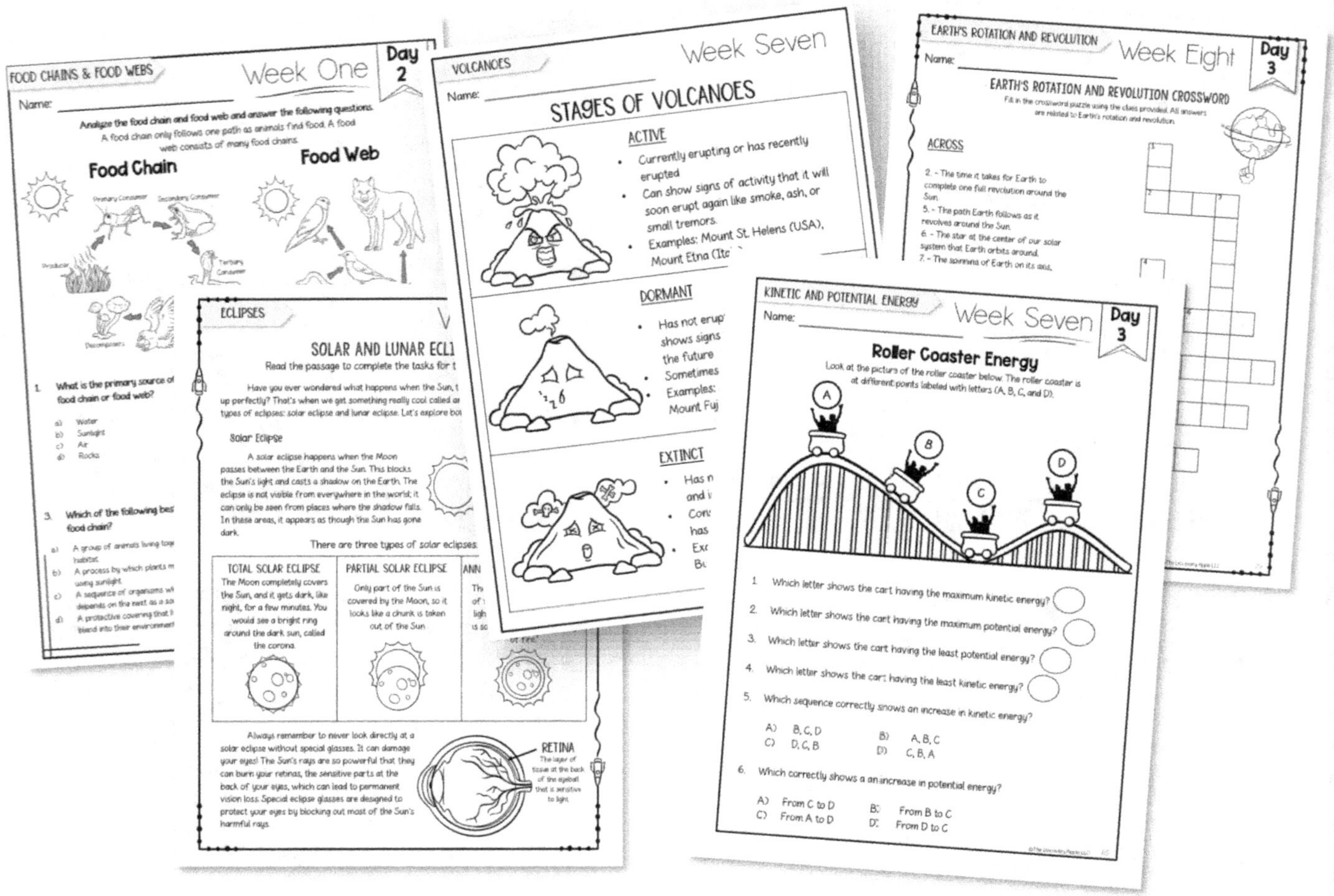

Author: Fatema Ghazi
Illustrator: Mariam Ghazi

Contact the author:
Fatema@thediscoveryapple.com

Workbook Highlights

WEEKLY TOPICS

Each week focuses on a different science topic, such as plant life cycle, animal adaptations, the human body, etc.

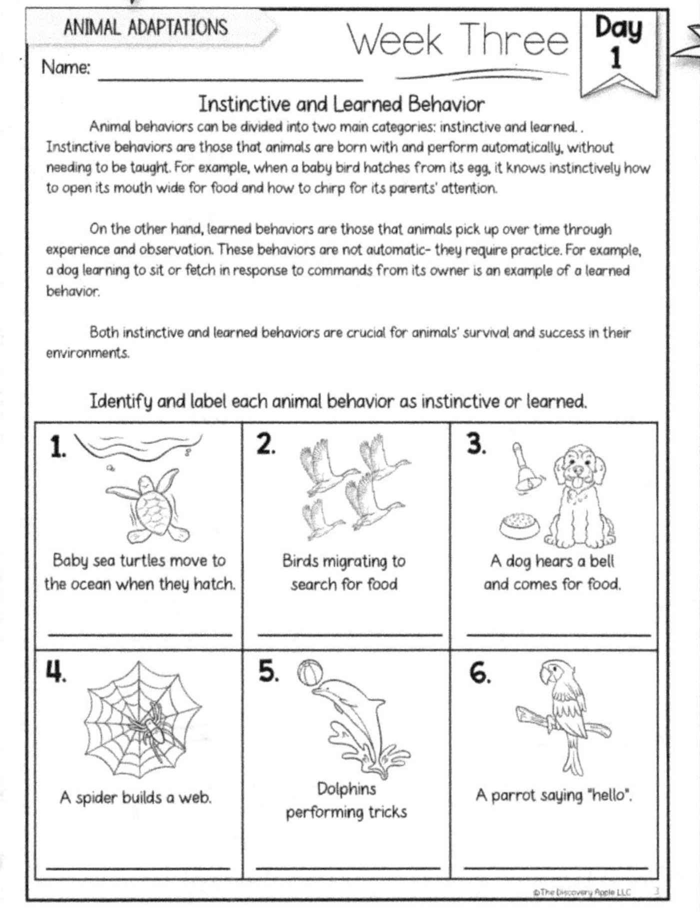

ANIMAL ADAPTATIONS

Week Three

Day 1

Name: ____________________

Instinctive and Learned Behavior

Animal behaviors can be divided into two main categories: instinctive and learned. . Instinctive behaviors are those that animals are born with and perform automatically, without needing to be taught. For example, when a baby bird hatches from its egg, it knows instinctively how to open its mouth wide for food and how to chirp for its parents' attention.

On the other hand, learned behaviors are those that animals pick up over time through experience and observation. These behaviors are not automatic- they require practice. For example, a dog learning to sit or fetch in response to commands from its owner is an example of a learned behavior.

Both instinctive and learned behaviors are crucial for animals' survival and success in their environments.

Identify and label each animal behavior as instinctive or learned.

1. Baby sea turtles move to the ocean when they hatch. ________	2. Birds migrating to search for food ________	3. A dog hears a bell and comes for food. ________
4. A spider builds a web. ________	5. Dolphins performing tricks ________	6. A parrot saying "hello". ________

©The Discovery Apple LLC 3

DAILY WORKSHEETS

Each topic consists of five worksheets, one for each weekday.

VARIETY OF ACTIVITIES

Each topic is explored through different types of activities:

- Reading Comprehension: Students read passages related to the topic and answer questions to demonstrate their understanding.
- Vocabulary: Activities focus on building and reinforcing vocabulary related to the topic, helping students grasp important scientific terms and concepts.
- Labeling Diagrams: Students practice identifying and labeling parts of diagrams or illustrations related to the topic, enhancing their understanding of its components.
- Critical Thinking Writing Response: Students are prompted to think critically about the topic and express their ideas through writing. These prompts encourage them to analyze, evaluate, and reflect on what they've learned, fostering higher-order thinking skills and creativity.

REVIEW AND REINFORCEMENT

Can be used as review and reinforcement of key concepts. You can use it alongside their regular science curriculum to deepen their understanding of daily life science topics.

FLEXIBLE USE

Can be used in various ways, such as for homework assignments, independent study, or as part of a classroom lesson. It can also serve as a tool for assessment and progress tracking.

Table of Contents

PHYSICAL SCIENCE

LIFE SCIENCE

Table of Contents

EARTH SCIENCE

SPACE SCIENCE

Exploring Physical Science

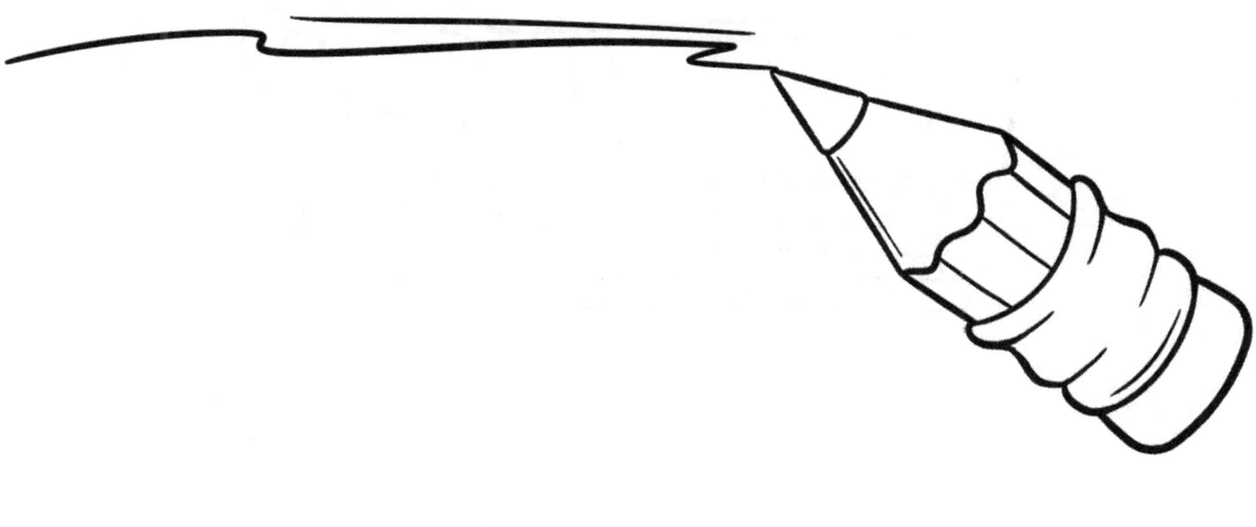

Week One

States of Matter

Matter is everywhere, in every place,
From the vast expanse of outer space,
To the book in front of your face.

Week One

Read the passage to complete the tasks for the next few days.

Matter is All Around Us

Take a look around you. What do you see? Everything you see is made of matter. The air you breathe is also made of matter, even though you cannot see it. The clothes you wear, even your own body, are made of matter. All of the planets, stars, and moons in the universe are made of matter. Matter is defined as anything that has mass and takes up space.

All matter is made of tiny particles called atoms. A group of atoms is called a molecule. Matter exists in several different forms depending on how its atoms are arranged. We call these forms states of matter. There are three main states of matter: solid, liquid, and gas.

Solids keep their own shape and do not flow because their molecules are tightly packed together. Picture a rock or a book on a shelf. They maintain their shape no matter where you put them. Other examples of solids are apples, computers, and the people around you.

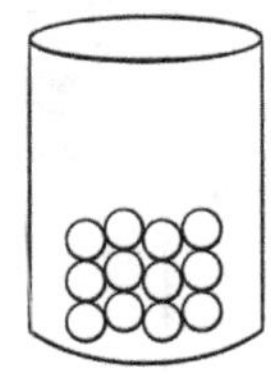

Molecules in a solid

Liquids do not have their own shape but can take the shape of the container they are in. For example, if you pour water into a cup, it will fill up and take the shape of the cup. If you pour that same water into a square container, it will be shaped like the container. Liquids are made up of molecules that are farther apart than in solids and can move around each other easily. Other examples of liquids are milk, juice, and oil.

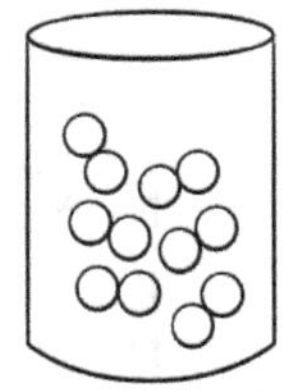

Molecules in a liquid

Gases are air-like substances that do not have a set size or shape. Their molecules drift around freely, so gases can expand to fill a large container or be squeezed into a smaller container. You can put your hand through gases and you will not feel them. Examples of gases include the helium in balloons, Earth's atmosphere, the air you are breathing, and water vapor.

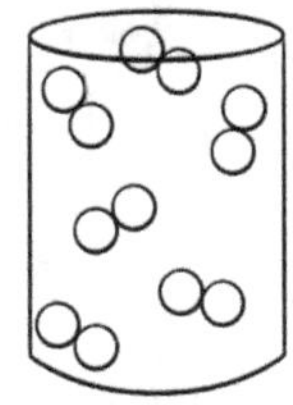

Molecules in a gas

Matter can change from one state to another. For example, water is a liquid, but if you freeze it, it turns into ice, which is a solid. If you heat the ice, it will turn into water vapor, which is a gas. Water vapor will change back into a liquid when cooled. Can you think of more examples of how matter can change from one state to another?

Week One

Day 1

Name: ______________________________

Solid, Liquid, or Gas

Look at each picture and write whether it shows a solid, liquid, or gas on the lines below.

1. ICE

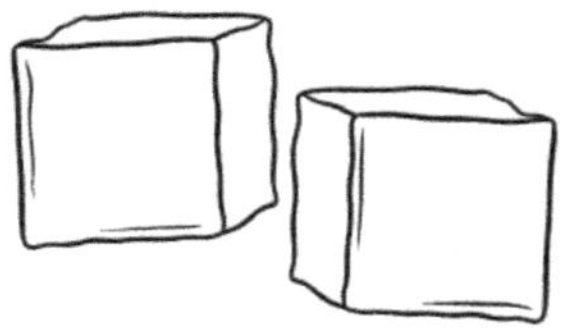

____

2. ORANGE JUICE

____

3. OLIVE OIL

____

4. OXYGEN

____

5. WOOD

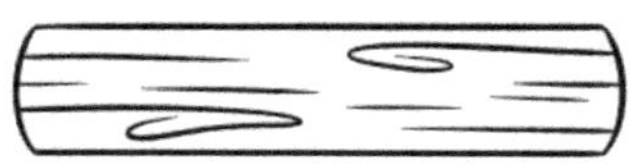

____

6. ROCKS

____

7. MILK

____

8. HELIUM IN BALLOONS

____

Name: ______________________________

Week One

Day 2

Solid, Liquid, or Gas

After reading the passage "Matter is All Around Us" on page 9, answer the following questions.

1. **Which state of matter takes the shape of its container?**
    a) Solid
    b) Liquid
    c) Gas
    d) None of the above

2. **What are the 3 states of matter?**
    a) Solid
    b) Liquid
    c) Gas
    d) All of the above

3. **Which state of matter has particles that are closely packed together?**
    a) Solid
    b) Liquid
    c) Gas
    d) Plasma

4. **Which of the following is an example of a gas?**
    a) Glass
    b) The air we breathe
    c) Syrup
    d) The food we eat

5. **What are the tiny particles that make up matter called?**

6. **What determines the shape of a liquid?**

7. **What is one example that shows how matter can change from one state to another?**

Name: ______________________

States of Matter

Match each word with its definition. Write the correct letter in the space provided.

1. Matter ________
2. Atom ________
3. Molecule ________
4. Solid ________
5. Liquid ________
6. Gas ________

(A) A state of matter that has a definite volume and definite shape

(B) A state of matter with no definite shape or volume

(C) Anything that has mass and takes up space (everything around you)

(D) A tiny particle that makes up matter

(E) A state of matter that has no definite shape but has a definite volume.

(F) A group of atoms bonded together

Under each set of molecules, write whether it is a solid, liquid, or gas.

7.
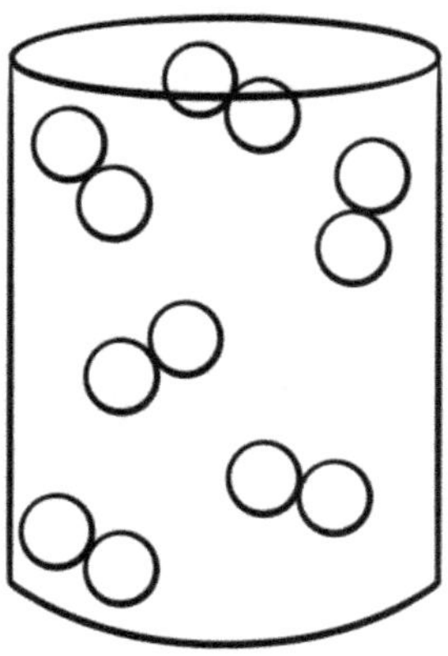

____

8.
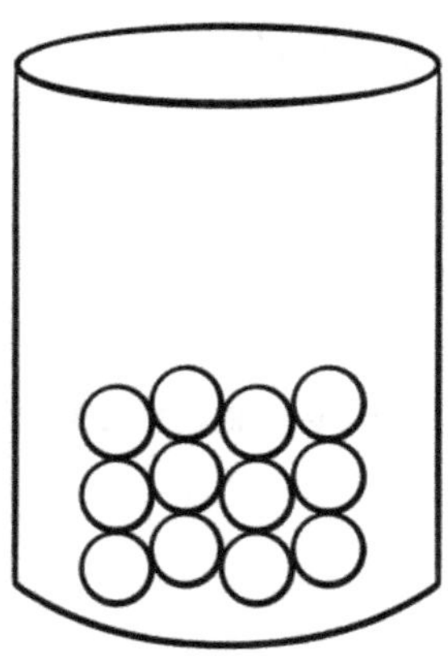

____

9.
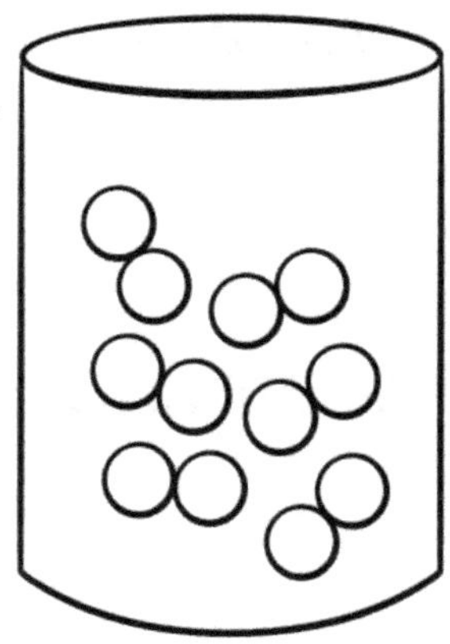

____

Name: ______________________

Week One

Day 4

Changes in Matter: Physical or Chemical

In the world around us, substances go through various transformations, often categorized as either physical or chemical changes.

Physical changes happen when something looks different or changes its shape, but it's still the same stuff. For example, when ice melts into water, it is still water, just not frozen anymore. When you tear paper into pieces, it is still paper, just smaller.

Chemical changes are when substances change into something new. For example, when metal rusts, it is not just metal anymore. It becomes a rusty material called rust. When vinegar and baking soda are mixed, they fizz and create a new substance that is different from either one.

After reading the passage, decide whether each of the described changes is a physical or chemical change. Write the letter 'P' for physical or 'C' for chemical under each image.

1. A nail gets rusty

______

2. Breaking glass

______

3. Chopping wood

______

4. An apple rotting

______

5. Peeling a banana

______

6. Frying an egg

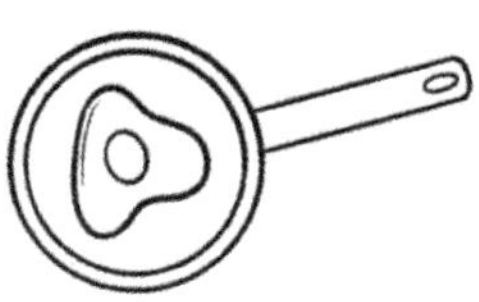

______

7. Burning wood

______

8. Fireworks exploding

______

9. Cutting paper

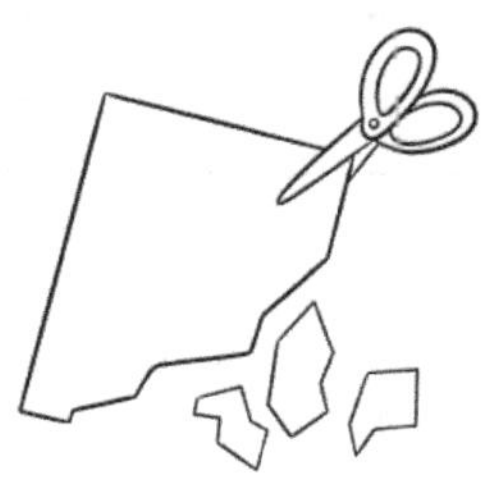

______

10. Ice cream melting

______

STATES OF MATTER

Week One

Day 5

Name: ______________________

Have you ever noticed how some things, like chocolate or candle wax, turn into a liquid when they get hot and then become solid again when they cool down? Why do you think this happens? Give three more examples of substances that do the same?

Week Two

Mixtures and Solutions

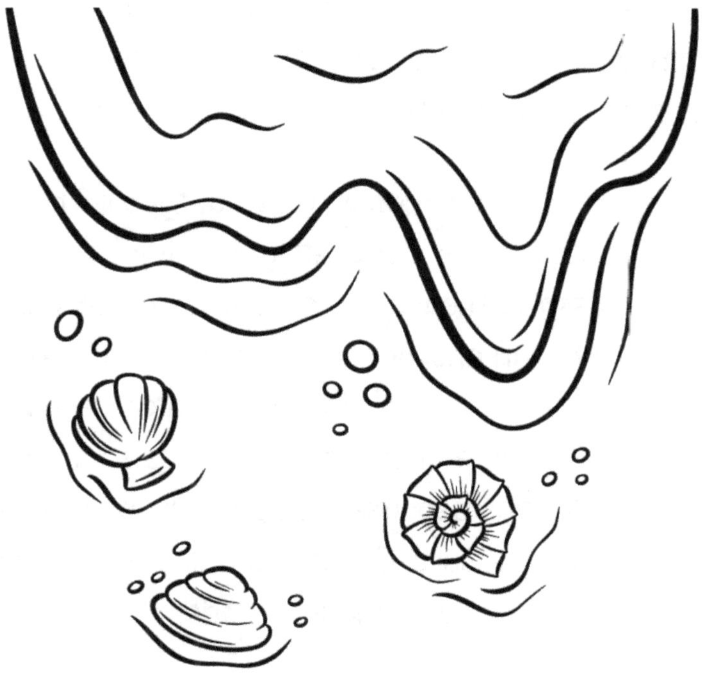

Mixtures can be found all around us, even in nature! Think about a sandy beach. The sand and water create a mixture, and if you look closely, you might find shells, rocks, and other treasures mixed in too.

Read the passage to complete the tasks for the next few days.

Understanding Mixtures and Solutions

Have you ever poured cereal in a bowl of milk? What happens to the cereal when you put it in the milk? The cereal basically floats around in the milk, right? Cereal and milk can be easily separated. This means that it is a **mixture**. A mixture happens when you combine two or more substances, but they do not completely mix. Each substance keeps its physical properties. They do not undergo a chemical change, such as something burning or rusting. Other examples of mixtures include a salad, sand mixed with water, and oil mixed with water.

A **solution** is a special type of mixture formed when one substance is completely dissolved in another substance. To dissolve means the substance breaks apart and spreads evenly through another substance until it is invisible. For example, you can't see the salt in ocean water because the salt is dissolved. Carbonated beverages, such as soda and energy drinks, are solutions because carbon dioxide gas is dissolved into liquid at high pressure. Other examples of solutions include dish soap, hot chocolate, and lemonade.

When something can dissolve into another substance, it has a physical property called **solubility**. Salt is considered soluble. If you put some salt in a bowl of soup and start stirring, the salt will spread through the soup until it is invisible, meaning it has dissolved. When a solid is dissolved in a liquid, we call the solid a **solute** and the liquid a **solvent**. Therefore, the salt is the solute and the soup is the solvent.

Name: ______________________

Week Two

Day 1

Mixture or Solution

Look at each picture and write whether it shows a **mixture** or a **solution** on the lines below.

1. SALAD

____

2. OIL AND WATER

____

3. LEMONADE

____

4. CHOCOLATE MILK

____

5. CEREAL AND MILK

____

6. LIQUID SOAP

____

7. SALT WATER

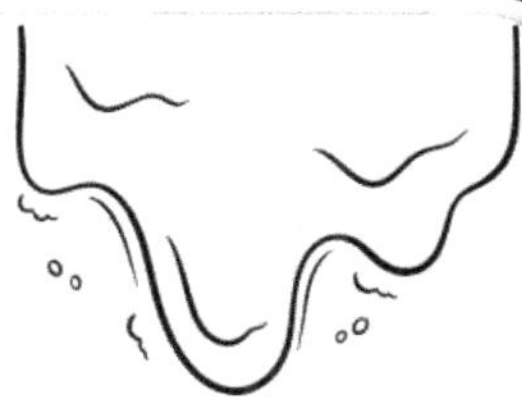

____

8. PIZZA

____

Name: ____________________

Week Two

Day 2

Understanding Mixtures and Solutions

After reading the passage "Understanding Mixtures and Solutions" on page 16, answer the following questions.

1. **What happens to a handful of sand when you put them in a cup of water? Is this an example of a mixture or solution?**
   - a) The sand will sink to the bottom of the cup. This is a solution.
   - b) The sand will sink to the bottom of the cup. This is a mixture.
   - c) The sand will dissolve. This is a mixture.
   - d) The sand will dissolve. This is a solution.

2. **When a solid is dissolved in a liquid, we call the liquid the __________.**
   - a) Solvent
   - b) Solute
   - c) Mixture
   - d) None of the above

3. **Sugar is an example of a __________ when it is dissolved in water.**
   - a) Solute
   - b) Solvent

4. **True or False: A mixture cannot be easily separated because it involves a chemical change.**
   - a) True
   - b) False

5. **Why can't you see salt in ocean water?**

6. **What are two examples of solutions?**

7. **Define a mixture in your own words.**

Name: ____________________________

Vocab Match

Match each word with its definition. Write the correct letter in the space provided.

1. Mixture ________
2. Solution ________
3. Solubility ________
4. Solvent ________
5. Solute ________
6. Dissolve ________
7. Physical Change ________

A. The substance that dissolves

B. The substance breaks apart and spreads evenly through a liquid until it is invisible

C. A special type of mixture where one substance is dissolved into another

D. A change in the shape or appearance of an object; no new substances are formed

E. The ability of something to dissolve in a liquid

F. A combination of two or more substances that keep their physical properties and can be easily separated

G. The liquid in which a solute is dissolved

Week Two

Day 4

Name: ____________________

Is it a mixture or a solution?

1. Not easily separated

   ____

2. Can be easily separated.

   ____

3. The solute dissolves into the solvent

   ____

4. Homogeneous (everything is evenly mixed)

   ____

5. Each substance keeps its physical properties

   ____

6. Heterogeneous (not evenly mixed)

   ____

Week Two

Day 5

Name: ______________________________

Think about the process of making lemonade. Write what ingredients are used to make lemonade and classify each as a solute or solvent. Explain what would happen if you added more sugar than the lemonade could dissolve.

Week Three

Force and Motion

Rockets rely on Newton's third law of motion to propel them into space. When the rocket engines push hot gases down, the gases push back on the rocket, making it go up.

Week Three

Read the passage to complete the tasks for the next few days.

Force and Motion

In the world around us, everything we see, feel, and experience involves force and motion. **Force** is simply a push or a pull exerted on an object. It can cause an object to start moving, stop moving, or change direction. Forces can be large or small, and they can act over short or long distances. For example, when we kick a soccer ball, we apply force to it, causing it to move. **Motion** refers to the change in position of an object over time. It can be described in terms of speed, direction, and acceleration (how an object speeds up or slows down).

FRICTION

Friction is the force that slows down objects when they rub against each other. Friction works against motion and acts in the opposite direction. For example, if you are running outside and you suddenly need to stop, you can stop because of the friction between your shoes and the ground.

GRAVITY

Gravity is a force of attraction. Gravity makes everything fall down toward Earth. The more mass an object has and the closer you are to the object, the stronger the gravity. Since Earth is huge, gravity is strong. Without Earth's gravity, we would fly right out of it; Earth would not be livable. If you kicked a ball, it would keep flying forever.

The Earth's gravitational force causes objects to accelerate, or speed up, as they fall. For example, if you drop a ball from a third-story window, the ball will keep falling and gain speed until it hits the ground. Gravity pulls the ball downward, causing it to move faster as it gets closer to Earth's surface.

MASS AND WEIGHT

Mass is the amount of matter in an object. Objects with more mass contain more matter. The mass of an object will never change. Your mass will be the same on Earth as in outer space. Weight is the force of gravity pulling on an object. Weight is measured in newtons, and the symbol for newtons is N. Unlike mass, the weight of an object can change depending on where it is. For example, you will weigh less on the Moon than you do on Earth because the Moon has weaker gravity. So, if you weigh 980 newtons on Earth, you will weigh about 162.2 newtons on the Moon.

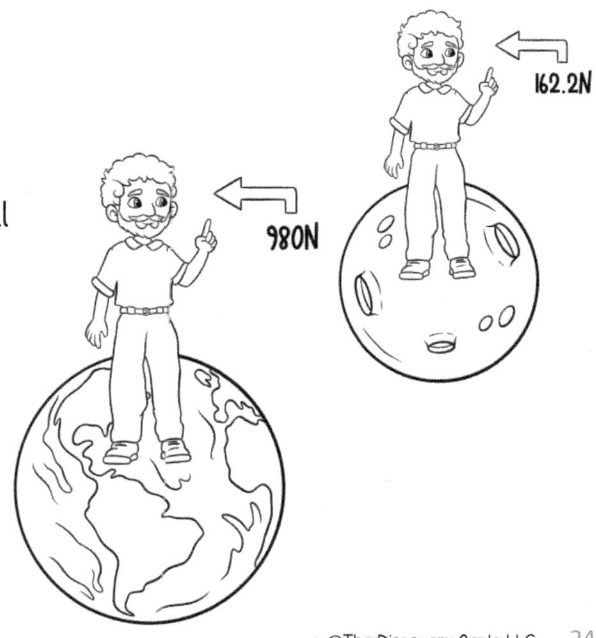

Name: ____________________

Week Three

Day 1

Force and Motion

1. **Force is:**

   a) The amount of matter in an object
   b) The speed of an object
   c) A push or pull on an object
   d) The distance an object travels

2. **Which force slows down or stops the motion of an object?**

   a) Friction
   b) Magnetism
   c) Gravity
   d) Air resistance

3. **What happens when you push a heavy box with more force?**

   a) The box becomes lighter
   b) The box moves faster
   c) The box becomes smaller
   d) The box stops moving

4. **Which force keeps objects on the ground?**

   a) Friction
   b) Magnetism
   c) Gravity
   d) Buoyancy

5. **How can force change the motion of an object? Give one example.**

____

____

____

____

6. **Explain how gravity affects objects on Earth.**

____

____

____

____

Read the passage to complete the tasks on the next page.

Newton's Laws of Motion

Isaac Newton is one of the most important scientists in history. Newton was a brilliant English mathematician, physicist, and astronomer who lived in the 17th and 18th centuries. He was best known for defining the three laws of motion and universal gravity.

1st Law of Motion

Newton's first law of motion states that an object at rest will remain at rest, and an object in motion will stay in motion with the same speed and direction unless acted on by an unbalanced force, such as a push or pull. For example, in the picture shown to the right, the soccer ball will not move until the girl kicks it.

2nd Law of Motion

Newton's second law of motion states that if you use the same force on two objects of different mass, you will get different accelerations (how objects speed up or slow down). The more force and the smaller the mass, the greater the acceleration. This relationship can be shown by the equation $F = ma$, where "F" represents force, "m" represents mass, and "a" represents acceleration.

For example, imagine pushing two different shopping carts, one empty and the other filled with groceries, with the same amount of force. According to Newton's second law, the shopping cart with the groceries will experience less acceleration compared to the empty one because it has more mass.

3rd Law of Motion

Newton's third law of motion states that for every action (force), there is an equal and opposite reaction (force).

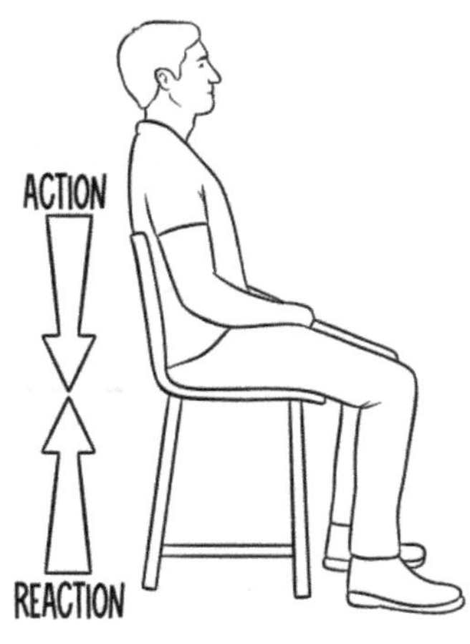

For example, when you sit on a chair, your body exerts a downward force on the chair (action) due to your weight. According to Newton's third law, the chair also exerts an equal and opposite force upward on your body (reaction). This upward force from the chair supports your weight and prevents you from falling through the chair.

Week Three

Day 2

Name: ______________________________

Newtons Laws of Motion

Directions: Decide which of Newton's Laws of Motion (First, Second, or Third) best describes each situation. Write "1st," "2nd," or "3rd" after each scenario to indicate your choice.

1. When you blow up a balloon and let it go, the air rushes out one way, and the balloon moves in the opposite direction.

____

2. A toy car continues rolling on a smooth surface until an unbalanced force causes it to stop.

____

3. A person needs to apply more force to push a heavy box compared to a lighter box..

____

4. When a person jumps off a diving board, the diving board pushes them upward into the air.

____

5. A paper airplane flies further when you throw it harder.

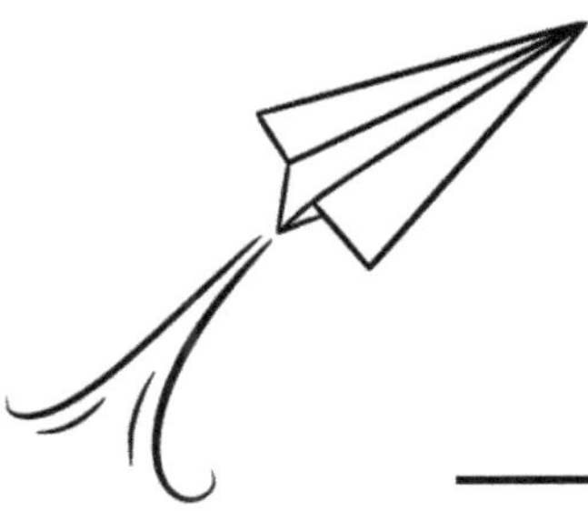

____

6. A skateboard slows down when you apply the brakes because of friction.

____

Week Three

Day 3

All About Friction

Try rubbing your hands together quickly. Do you feel the heat? If not, try increasing the pressure as you rub. When you rub your hands together, you generate heat because of friction. The faster and harder you rub, the more friction is produced, resulting in an increase in temperature.

Friction is a force that resists motion when two objects rub against each other. Anytime two objects rub against each other, friction occurs. It works against the motion and acts in the opposite direction.

The amount of friction depends on the materials from which the two surfaces are made. For example, rough surfaces like rock and sand create more friction than smooth surfaces like glass and ice. It is harder to move things when there is a lot of friction holding them back. For example, you would have to push a box using more force to get it moving on a carpet than you would on a tile floor. There is more friction between the carpet and the box than between the tile and the box.

Did you know that if friction did not exist, you would slip and slide around wherever you take a step? The friction between your shoes and the floor stops you from slipping. It would also be very tough to stand up without friction.

There are three different types of friction: dry friction, fluid friction, and rolling friction. **Dry friction** occurs when two solid objects touch each other. If they are not moving, it is called static friction. If they are moving, it is called kinetic or sliding friction. **Fluid friction** involves liquid or air moving around an object. A submarine moving through water is an example of fluid friction. Lastly, **rolling friction** occurs when a round object rolls over a surface. For example, a bowling ball or truck tires have rolling friction.

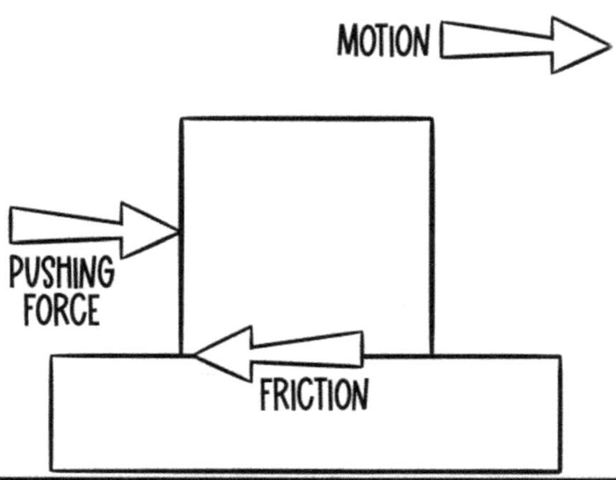

Name: ______________________________

Recognizing Push, Pull, Friction, and Gravity

Identify whether each example demonstrates a push, a pull, friction, or gravity.

1. Opening your car door

____

2. Rubbing your hands together to generate warmth

____

3.

Picking a flower

____

4.

Falling back to the ground after jumping in the air

____

5. Moving a shopping cart

____

6.

An apple falling off a tree

____

7. Gripping a pencil tightly to prevent it from slipping while writing

____

8.

Brakes on a car slowing down the vehicle

____

Name: ____________________

Week Three

Day 4

Vocab Match

Match each word with its definition. Write the correct letter in the space provided.

1. Force ________
2. Motion ________
3. Friction ________
4. Gravity ________
5. Mass ________
6. Weight ________
7. Acceleration ________
8. Speed ________
9. Velocity ________
10. Inertia ________

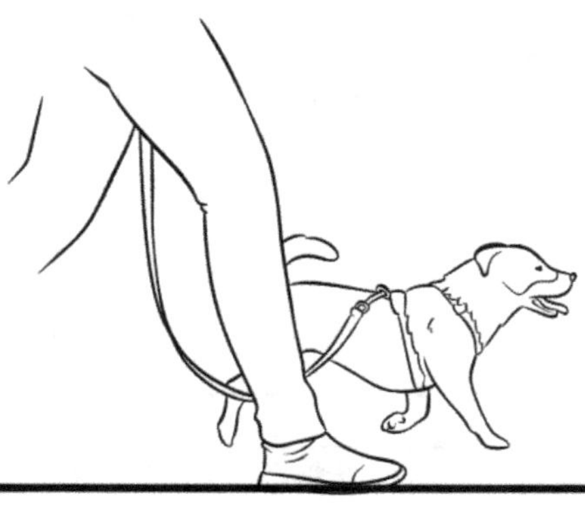

(A) The force that pulls objects toward each other

(B) A push or a pull on an object

(C) The speed and direction of a moving object

(D) The rate at which an object speeds up, slows down, or changes direction

(E) The force that slows down objects when they rub against each other

(F) The amount of matter ("stuff") in an object

(G) An object's tendency to resist a change in motion

(H) The change in position of an object

(I) The distance an object travels in one unit of time

(J) The amount of force that gravity has on an object's mass

Week Three

Day 5

Name: ______________________________

How does the type of surface affect the amount of friction experienced by an object? Consider the differences in friction between rough and smooth surfaces and give examples to support your answer.

Week Four

Simple Machines

Wheel and axles make it easier to move heavy objects by reducing friction. Think about how much easier it is to roll a suitcase with wheels compared to carrying it!

Week Four

Read the passage to complete the tasks for the next few days.

Simple Machines

Simple machines are everywhere! Did you ever ride a bike or climb up a ladder? Did you use a hammer to build something or use scissors to cut paper? These are all different types of simple machines that people use every day. Simple machines are basically tools that people use to make work easier. For example, scissors make it easier to cut out shapes.

Simple machines can be broken down into six different categories: the lever, the wheel and axle, the pulley, the inclined plane, the wedge, and the screw.

A **lever** is a bar that rests on a fulcrum and can be used to lift or move loads. The fulcrum is the point on which the lever turns or balances. Some examples of levers are wheelbarrows, scissors, and door handles.

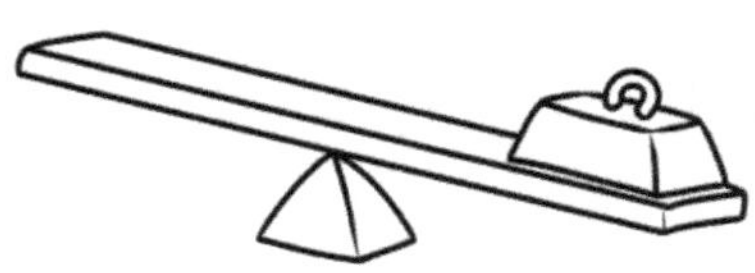

The **wheel and axle** is basically a wheel combined with a rod going through its center as an axle. It can be used to lift or move loads. An electric fan, a clock, and anything with a wheel, such as bikes, cars, and skates, are examples of a wheel and axle.

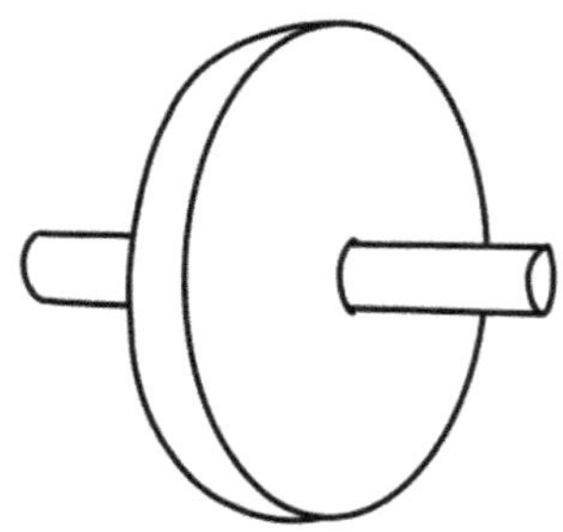

A **pulley** is a type of wheel and axle that is combined with a rope or a chain. The wheel includes a groove along the edges to guide the rope and raise, lower, or move a load back and forth. Examples of pulleys include a crane, window blinds, and elevators.

Week Four

An **inclined plane** is any slanted surface where one end is higher than the opposite end. Many people call an inclined plane a ramp. It has no moving parts. Inclined planes help move heavy objects up and down heights. Examples of inclined planes that you might see every day are wheelchair ramps, slides, and stairs.

A **wedge** is made up of two inclined planes joined together. The planes meet and form a sharp edge. You can find wedges made of wood, plastic, metal, or even stone. The wedge is used to separate objects or keep things together. For example, scissors are used for separating things and door stoppers are used to hold the door in place.

An inclined plane wrapped around a pole is called a **screw**. It usually has a flat end and a sharp-pointed end. A screw can be used to hold things together, tighten things, lift heavy objects, and drill holes in objects.

Look around you. What simple machines can you find?

Week Four

Day 1

Directions: Write the simple machine that each picture represents. Some pictures will include two simple machines working together, also known as a compound machine.

Label It!

WORD BANK			
	Lever	Screw	Wheel & Axle
	Wedge	Inclined Plane	Pulley

1. 

____

2. 

____

____

3. 

____

4. 

____

5. 

____

6. 

____

7. 

____

8. 

____

9. 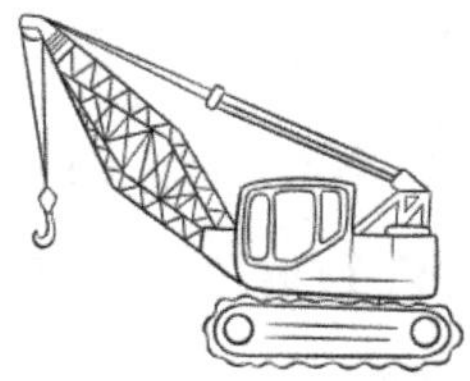

____

____

10. 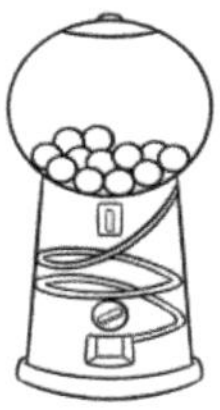

____

____

11. 

____

____

12. 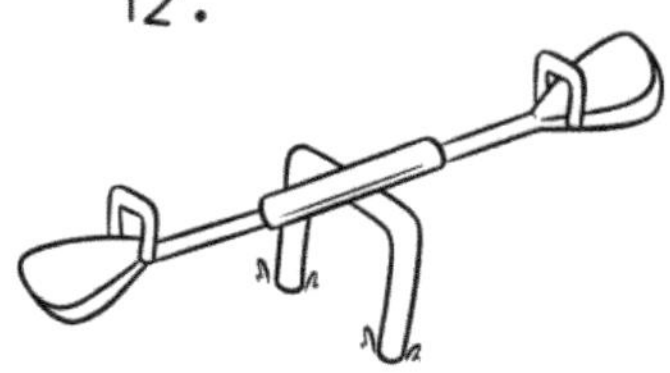

____

Name: ______________________________

Week Four

Day 2

Simple Machines

1. **What is the name of the point or object that a lever rests on?**
   a) Fulcrum
   b) Wedge
   c) Gear
   d) Spring

2. **What do people use pulleys for?**
   a) To cut through object
   b) To drive vehicles
   c) To lift or move heavy objects
   d) To measure distances

3. **What do people use wedges for?**
   a) To measure objects
   b) To separate an object or keep things together
   c) To lift or more heavy objects
   d) To climb up objects

4. **What is a compound machine?**
   a) A machine that uses electricity as its power source
   b) A machine designed for simple tasks
   c) A machine used for transportation purposes
   d) Two or more machines working together

5. **List two objects that use a screw.**

6. **What simple machine helps move heavy objects up and down heights?**

7. **What simple machine is made up of two inclined planes joined together?**

Name: ______________________

Week Four

Day 3

Vocab Match

Match each word with its definition. Write the correct letter in the space provided.

1. Simple Machines ________
2. Compound Machines ________
3. Fulcrum ________
4. Work ________
5. Load ________
6. Force ________
7. Pulley ________
8. Lever ________
9. Inclined Plane ________
10. Wheel & Axle ________
11. Screw ________
12. Wedge ________

A. A slope or ramp that makes it easier to move a load up or down

B. What is being moved

C. Machines made up of two or more simple machines that work together

D. A straight, rigid object that rests on a fulcrum and lifts or moves loads

E. The point on which the lever turns or balances

F. A push or a pull

G. An inclined plane wrapped around a cylinder that helps hold things together

H. Tools with few or no moving parts that make work easier

I. Made up of two inclined planes that can be used to push objects apart or move them together

J. A wheel with a rod called an axle through the center that helps move a load

K. Using force to move a load

L. A wheel with a groove and a rope used to lift or move objects

Name: ____________________

Simple Machines Scavenger Hunt

Find examples of each simple machine. List at least 3 examples in the correct box.

Ready, Set, Go!

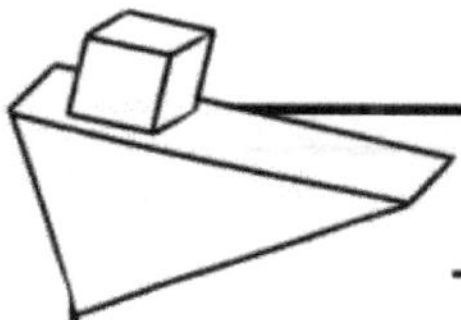

INCLINED PLANE

LEVER

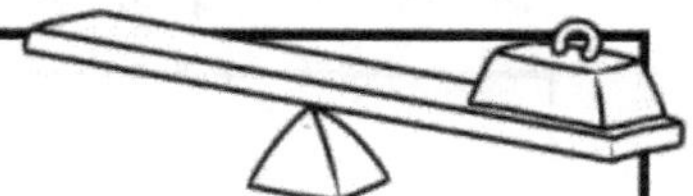

SCREW

PULLEY

WEDGE

WHEEL AND AXLE

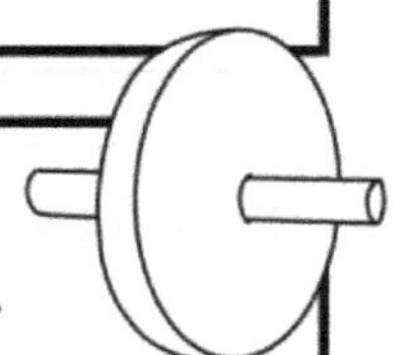

Name: ______________________________

Week Four

Day 5

Hello Inventor! You are a smart person, and I know you can come up with amazing ideas. Think of something that will help a person accomplish a task. A task is work that needs to be done. Think about the different types of simple machines you have learned about. Your invention should include two or more simple machines working together. YOU CAN DO THIS!

1. Brainstorm your ideas / Rough draft

2. Explain what simple machines are needed for you invention to work.

____

____

____

____

____

Name: ______________________________

Week Four

Day 5

3. What is the specific task your invention aims to assist with? Explain how your invention will make the person's task easier.

____

____

____

____

____

____

____

____

____

4. Draw and color your invention. Then, label the simple machines in your drawing.

Week Five

Electricity

Circuits, Conductors, and Insulators

Ever get a shock when you touch a metal doorknob after rubbing your feet on carpet? That's because you built up static electricity, and when you touched the metal, the electrons jumped from your body to the metal, causing a tiny zap!

Week Five

Read the passage to complete the tasks for the next two days.

Series and Parallel Circuits

A series circuit and a parallel circuit are the two basic types of electric circuits that can be found in electrical devices. In a **series circuit**, electricity has only one path to follow. All the parts are connected one after another with no branches. From one end of the battery (cell), electricity will move along one path through the resistors to the other end of the battery. A **resistor** is a part of a circuit that uses some of the electrical energy to do work, such as producing light. An example of a resistor in a circuit is a light bulb.

SERIES CIRCUIT

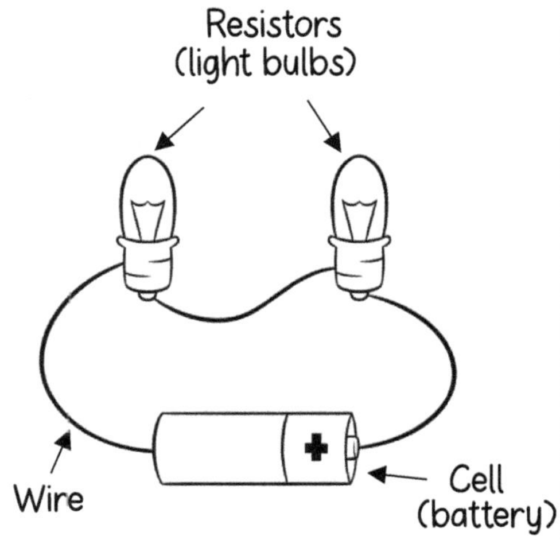

If you take out or damage any bulb in a series circuit, the entire circuit will not function. This is because the path is cut off and electricity cannot flow. An example of a series circuit is a string of Christmas lights. A disadvantage of this type of circuit is that one burned-out bulb causes all of the lights to go out. Also, the more bulbs you add to the circuit, the dimmer the lights become. This happens because the battery's power is shared among all the bulbs, so each bulb gets less energy. One advantage of a series circuit is that you will always know if there is a break because everything stops working.

PARALLEL CIRCUIT

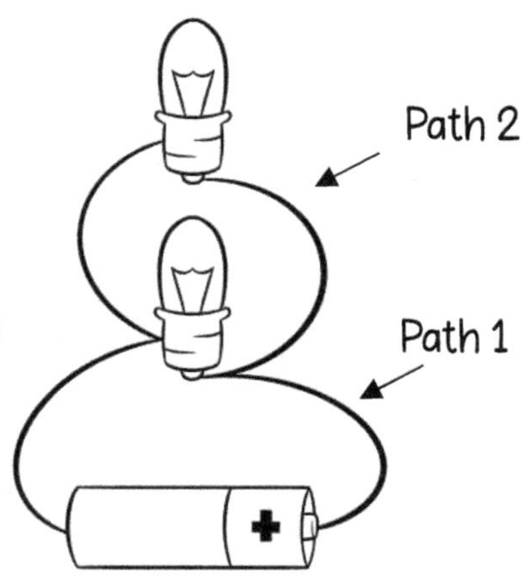

In a **parallel circuit**, there are multiple pathways. Since electricity has more than one route to take, the circuit will still work even if one path fails. This is why parallel circuits are more common and are found all around us, such as in classrooms and homes. Think of the light bulb in your bedroom. If it burns out, the light bulb in the kitchen and the rest of the house are not affected. Unlike series circuits, parallel circuits allow lights to stay bright no matter how many bulbs are added. The main benefit of a parallel circuit is that everything can keep working, even if one component fails.

Name: ______________________

Week Five

Day 1

Series and Parallel Circuits

1. Explain the difference between a series circuit and a parallel circuit. Include how electricity flows and what happens if one bulb goes out.

____

____

____

____

____

____

____

____

____

Identify and write the type of circuit shown. Is it a series or a parallel circuit?

2.

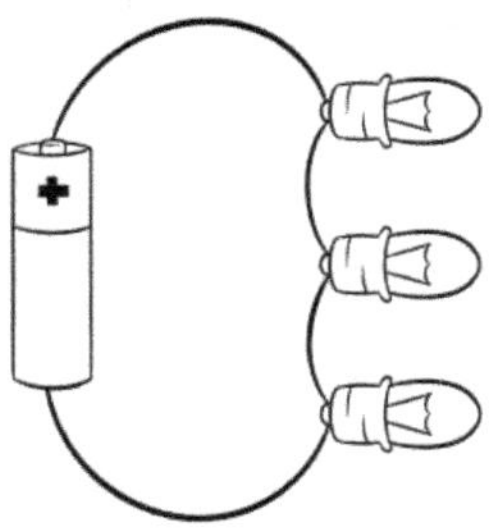

____

3.

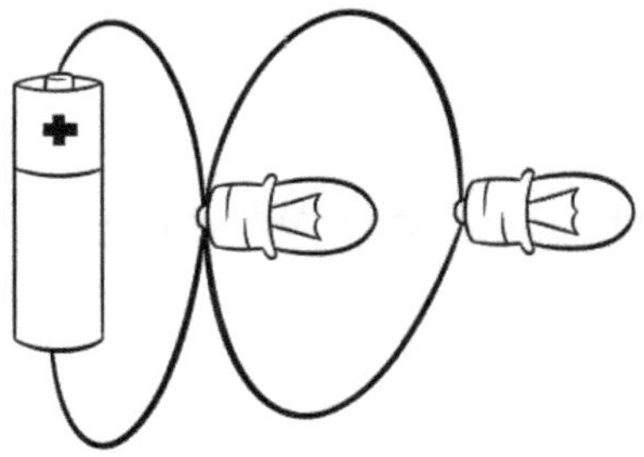

____

4.

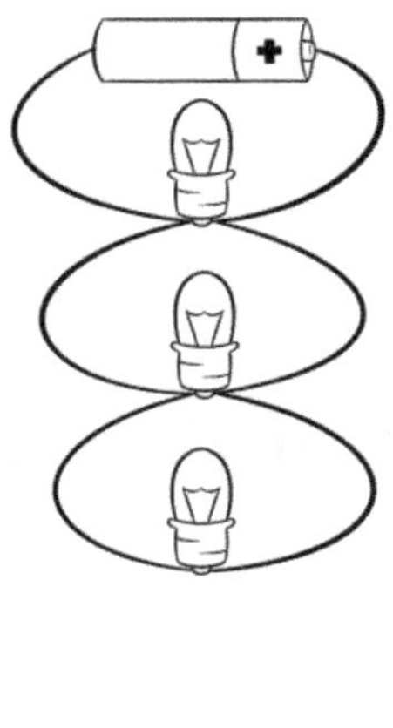

____

5.

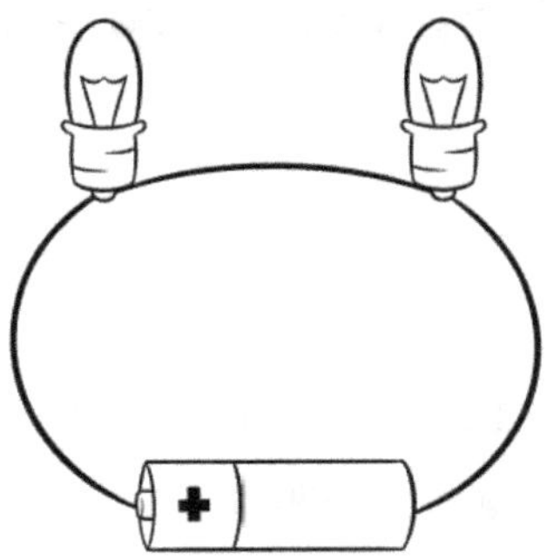

____

Series and Parallel Circuits

1. What type of circuit has multiple pathways?

   a) Series
   b) Parallel
   c) Both
   d) None

2. What is a battery also known as?

   a) Electrons
   b) Resistors
   c) Circuit
   d) Cell

3. If a parallel circuit has two light bulbs and you add two more bulbs, will the lights get dimmer?

   a) Yes
   b) No

4. Does this picture show a series circuit or a parallel circuit?

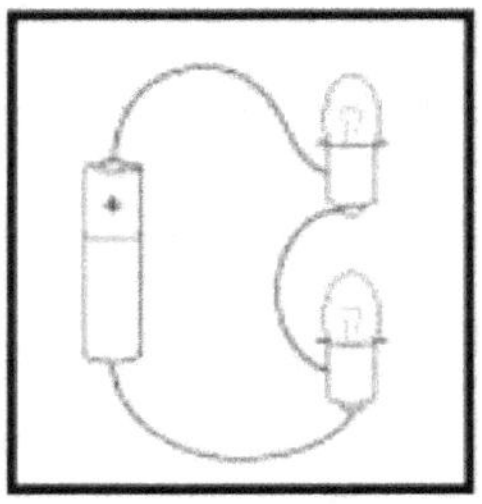

   a) Series
   b) Parallel
   c) None

5. What is a resistor? What is an example of a resistor.

6. What is one disadvantage of a series circuit?

7. If a light bulb in your home bathroom burns out, will the lights in the kitchen be affected? Explain your reason.

Week Five

Day 3

Conductors and Insulators

Conductors are materials that make it easy for electricity to move through them. Metals like copper are great conductors because they let electricity flow without much resistance. Gold and silver are other examples of conductors. Did you know water can also conduct electricity, especially when it contains minerals or chemicals? That's why it's not safe to have electrical things near water, like in a pool.

Insulators are like shields that slow down or block electricity from passing through. Unlike conductors, they don't let electricity flow easily. Materials like rubber, plastic, glass, and ceramics are good insulators. They're used to cover wires to keep electricity contained and prevent accidents.

Directions: Identify the conductors and insulators.
Record your answer under each picture. Write "C" for conductor and "I" for insulator.

1. Aluminum Cans  ________	**2.** Gold Ring  ________	**3.** Cotton Shirt  ________	**4.** Rubber Band 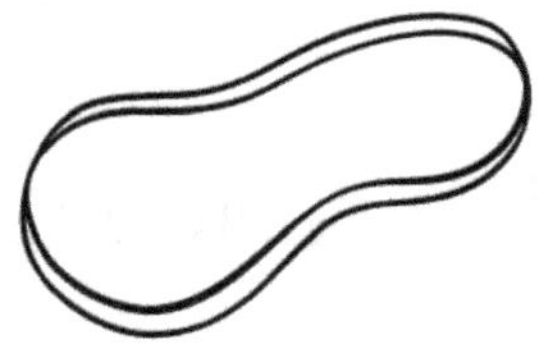 ________
5. Your hand 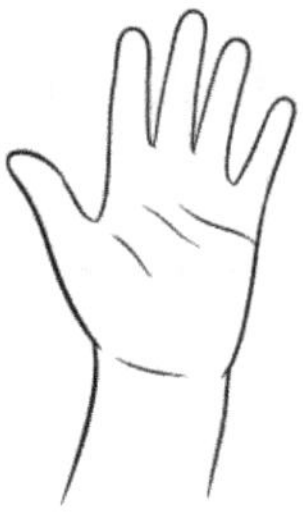 ________	**6.** Glass marbles  ________	**7.** Silver Spoon  ________	**8.** Notebook Paper 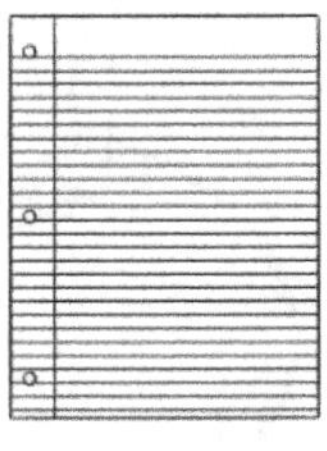 ________

Vocab Match

Match each word with its definition. Write the correct letter in the space provided.

1. Open Circuit ________
2. Closed Circuit ________
3. Series Circuit ________
4. Parallel Circuit ________
5. Electric Current ________
6. Switch ________
7. Conductors ________
8. Insulators ________
9. Electron ________

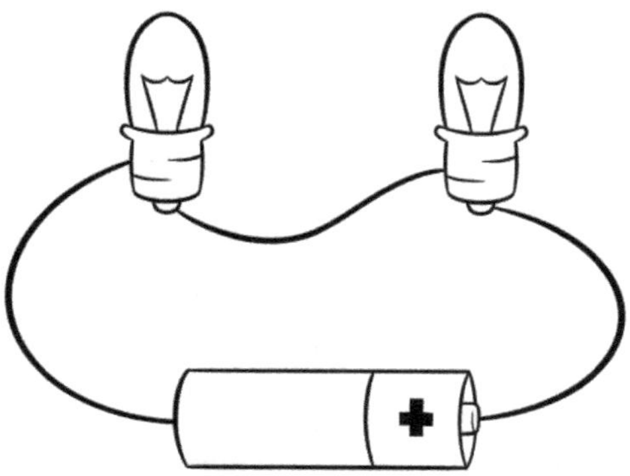

A. Materials that allow electricity to pass through them easily

B. A circuit that is complete and where current flows

C. A device that can complete or break the connection of an electric circuit

D. Materials that do not allow electricity to pass through them

E. A circuit in which the electric current follows two or more paths

F. A particle with a negative charge

G. A circuit that is incomplete and where current does not flow

H. The flow of electricity through a circuit

I. A circuit in which electricity flows through only one path

Week Five

Day 5

Pretend you are an inventor creating a new electrical device for your home. Explain how your device works. Be sure to include the type of circuit it uses and why you chose certain conductors or insulators.

Week Five

Day 5

Name your electrical device and illustrate it in the box below.

Name of Device: ______________________________

Week Six

Light and (Heat) Thermal Energy

Did you know that some snakes have a special way of seeing called infrared vision? It helps them find food and move around at night by feeling the heat from other animals.

Week Six

Read the passage and answer the questions on the next page.

Reflection, Refraction, and Absorption

With the help of light, you can see everything around you. Without light, nothing would be visible. Light travels in a straight path until it hits an object. When light hits an object, it can be absorbed, reflected, or refracted.

REFLECTION

Reflection is when energy in the form of light bounces off the surface of an object. For example, when you see a yellow banana, it is because the banana reflects yellow light into your eyes. When you look in a mirror, the image you see is a reflection of yourself. Light bounces off the mirror and enters your eye.

REFRACTION

Refraction is the bending of light. When light moves from one medium (like air) to another medium (like water), it changes direction, causing the light to bend. For example, when you put a pencil in a cup of water, it looks like the pencil is bent. This happens because the light moves from water to air, causing it to refract and making the underwater part of the pencil to appear in a different place than it actually is.

ABSORPTION

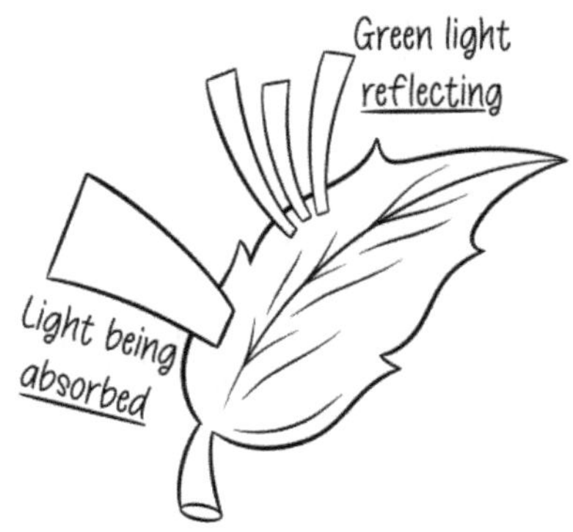

Absorption is when light is taken in by an object and does not reflect or refract. The light energy that hits an object is converted into heat energy. The color black absorbs more light and feels hotter than white objects. If an object absorbs all colors except one, we see the color it does not absorb. For example, a leaf absorbs all colors except green light, which is why leaves appear green.

Reflection, Refraction, and Absorption

1. **When you see your reflection in a mirror, what type of process is occurring?**
   a) Absorption
   b) Reflection
   c) Refraction
   d) Diffraction

2. **Which of the following best describes refraction?**
   a) Light being absorbed by an object
   b) Light bouncing off the surface of an object
   c) Light reflecting off a smooth surface
   d) Light bending as it passes through different mediums

3. **Which of the following is an example of absorption of light?**
   a) Light bouncing off a mirror
   b) Light passing through a glass window
   c) Light being taken in by a black shirt
   d) Light refracting through a prism

4. **What happens to light when it strikes a smooth surface at an angle and bounces off at an equal angle?**
   a) Reflection
   b) Absorption
   c) Refraction
   d) Scattering

Determine whether each picture is showing reflection, refraction, or absorption.

5.

____

6.

____

7.

____

8.

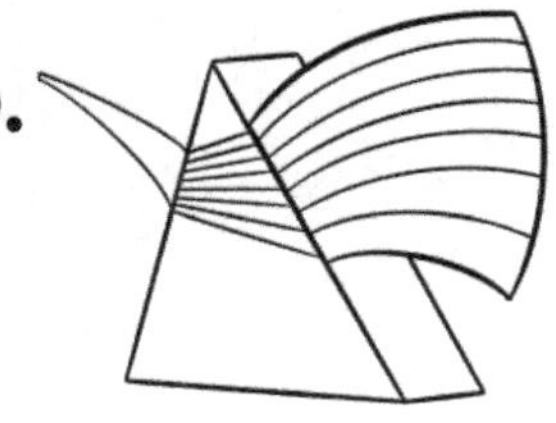

____

9.

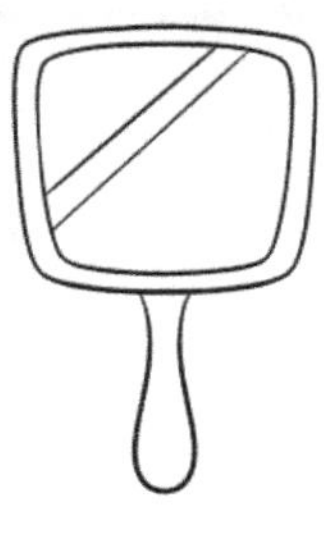

____

10.

____

Transparent, Translucent, and Opaque

Transparent: When something is transparent, you can see through it clearly. Light passes through it without being scattered or blocked. Examples include clean glass and clear plastic.

Translucent: When something is translucent, light can pass through it, but you cannot see through it clearly. It scatters the light as it passes through, making images look blurry. Examples include frosted glass and thin paper.

Opaque: When something is opaque, you cannot see through it at all. Light does not pass through it. Examples include wood and metal.

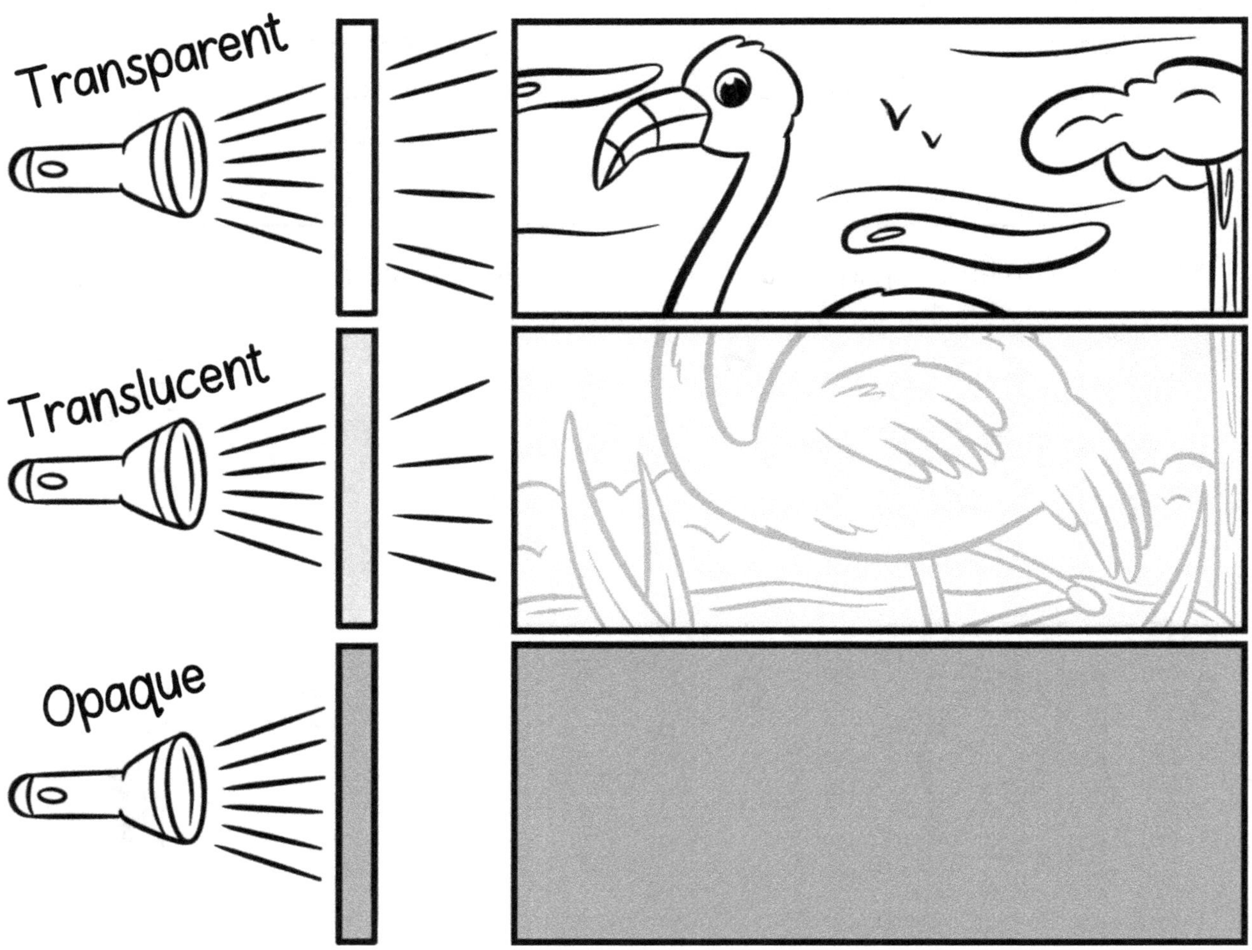

Week Six

Day 2

Transparent, Translucent, or Opaque

Does the image show a transparent, translucent, or an opaque object?

1. SUNGLASSES

____

2. CLEAR GLASS WINDOW

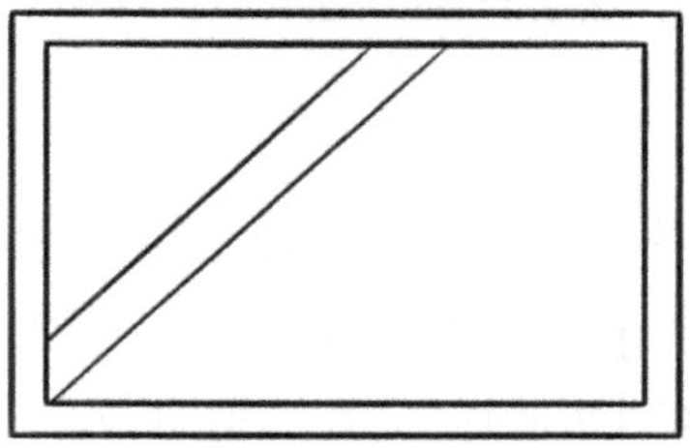

____

3. INFLATED BALLOONS

____

4. APPLE

____

5. BRICK WALL

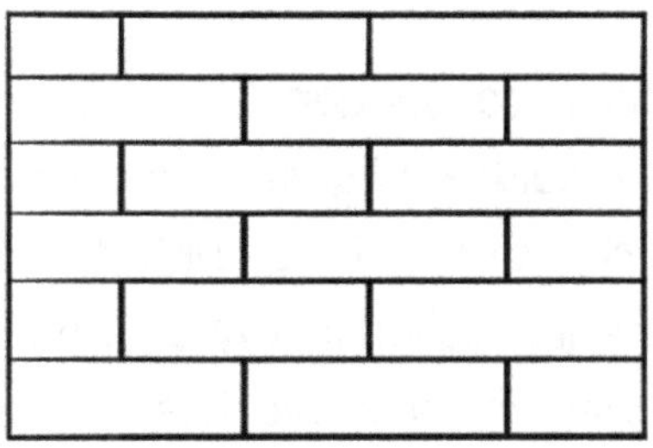

____

6. WAX PAPER

____

7. EYEGLASSES

____

8. BOOKS

____

9. CLEAR WATER IN A GLASS

____

Heat Transfer

Conduction, Convection, and Radiation

Did you know that heat is always moving? Heat always moves from a warmer object to a cooler object. Cold does not move on its own. For example, when you hold a bowl of hot soup in your hands, heat travels from the hot soup to your hands. The transfer of heat continues until both objects reach the same temperature. Heat can transfer (move) in three ways: conduction, convection, and radiation.

CONDUCTION

Conduction is the transfer of heat through direct contact between two objects that are touching. Imagine touching a hot stove with your hand. Ouch! Your hand will burn because the hot stove transfers heat energy to your hand. This is an example of conduction. Some materials are better conductors than others, meaning they allow heat to pass more easily. Metal objects get hot quickly because they are good conductors. Plastic and wood take longer to heat up, so they are poor conductors.

CONVECTION

Another type of heat transfer is **convection**. Convection occurs when heat transfers through fluids. Liquids and gases are considered fluids. When fluids are heated, they rise. Cooler fluids sink down. This continuous rising and sinking movement is called a **convection current**. For example, when you heat a pot of water on the stove, the water at the bottom heats up first and rises. The cooler water sinks to the bottom, where it is heated. Convection currents also explain why the air near the floor is cooler than the air near the ceiling.

RADIATION

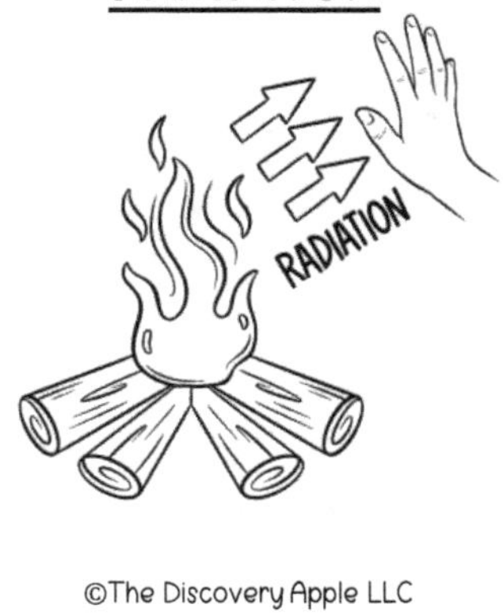

The third type of heat transfer is **radiation**. Radiation is the transfer of heat through empty space using electromagnetic waves. The largest source of radiation is the sun, which sends heat to Earth through electromagnetic waves. Radiation allows you to feel heat without touching the source. For example, when you hold your hands near a fire, you can feel the heat even though you are not touching it. The heat travels through space until it reaches your hands.

Heat Transfer

Conduction, Convection, and Radiation

1. Is a metal spoon a good conductor or a bad conductor?

   a) Good conductor
   b) Bad conductor

2. You raised your hand above a tea pot and felt the heat. This is an example of ________.

   a) Conduction
   b) Convection
   c) Radiation

3. ____________ is when heat transfers through liquids and gases.

   a) Conduction
   b) Convection
   c) Radiation

4. What is the biggest source of radiation?

   a) The Sun
   b) Fire
   c) A stove
   d) Hot water

5. When you heat up a pot of soup, the heated liquid ____________________________,
   (Rises up / sinks down)
   and the cooler liquid ____________________________.
   (Rises up / sinks down)

6. Draw an example of conduction, convection, and radiation in the boxes below.

Name: ______________________

Week Six

Day 4

Vocab Match

Match each word with its definition. Write the correct letter in the space provided.

1. Light Energy ________
2. Thermal Energy ________
3. Reflection ________
4. Refraction ________
5. Absorption ________
6. Transparent ________
7. Translucent ________
8. Opaque ________
9. Conduction ________
10. Convection ________
11. Radiation ________

A. When light is taken in by an object and does not reflect or refract

B. Transfer of heat through direct contact with two objects that are touching

C. A form of electromagnetic energy that travels as a wave and can move through empty space, such as visible light

D. When energy in the form of light bounces off the surface of an object

E. Transfer of heat through space on electromagnetic waves

F. Some light passes through

G. All light passes through

H. Transfer of heat through fluids

I. Energy from moving particles, also known as heat energy

J. No light passes through

K. The bending of light

Week Six

Day 5

Think about your daily life. Write about two examples of reflection and two examples of refraction that you see around you. Explain what happens to the light in each example.

Week Seven

Kinetic and Potential Energy

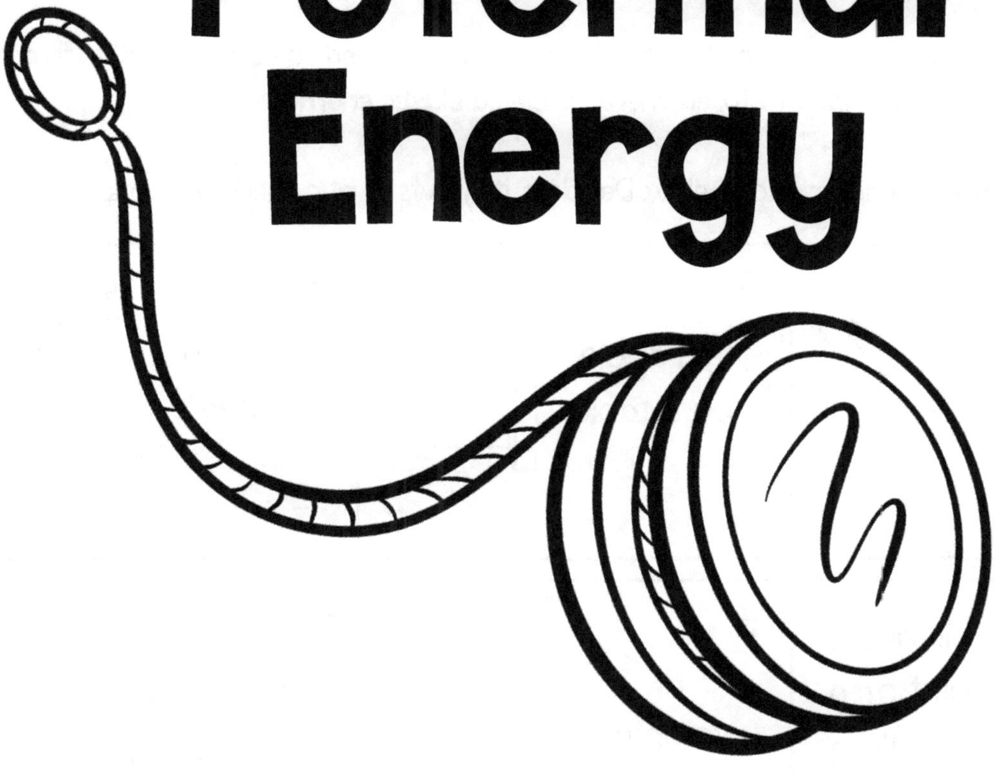

Did you know? A yo-yo is a fun example of potential and kinetic energy in action! When you pull the yo-yo up, you're giving it potential energy. As it falls back down, that potential energy is converted into kinetic energy, making the yo-yo spin.

Energy in Action

Energy is the ability to do work. We need energy for everything we do. Energy is all around us and comes in many different forms.

Stand up and jump 10 times. Are you tired? Take a short break, then jump again. Did you know you were changing forms of energy? When you were jumping, you were using **kinetic energy,** which is energy of motion. When you took a break, you had **potential energy,** which is energy that is stored and not being used yet.

Some potential energy is stored in objects that are stretched or compressed, such as a slingshot or a spring. This type of stored energy is called **elastic energy**.

Look at each picture and description below. Decide if it shows kinetic energy or potential energy.

1. A paper airplane flying  ____________	2. An apple on a table  ____________	3. A fan turned on  ____________
4. 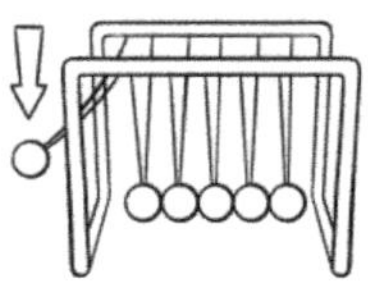 Highest point of a pendulum ____________	5. A waterfall  ____________	6. Pushing a wheelbarrow  ____________
7. A pulled slingshot 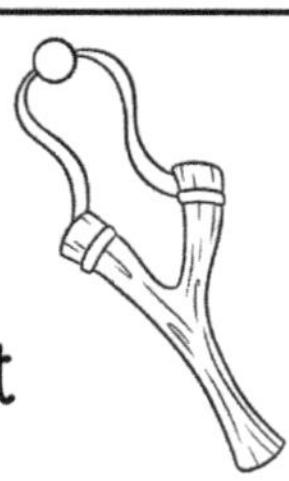 ____________	8. A moving car  ____________	9. A parked car  ____________

Energy in Action

1. **A moving object always has**

   a) Kinetic energy
   b) Potential energy

2. **Which ball has more potential energy?**

   a) A ball sitting at the top of a hill
   b) A ball sitting at the bottom of a hill
   c) A ball rolling down a hill
   d) A ball bouncing up and down on a flat surface

3. **If an object is not moving, it has**

   a) Kinetic energy
   b) Potential energy

4. **Which car has more kinetic energy?**

   a) A car parked on the side of the road
   b) A car driving 30 miles per hour
   c) A car driving at 60 miles per hour
   d) None of the above

5. **Which spring has the most potential energy?**

   a) Spring 1
   b) Spring 2

6. **Can you think of a game or activity that uses both kinetic and potential energy? Explain how each type of energy is used.**

Name: ______________________________

Roller Coaster Energy

Read the paragraph below and view the diagram on the next page to answer the questions that follow.

When you ride a roller coaster, you experience two types of energy: potential energy and kinetic energy.

Potential energy is the energy an object has because of its position. The higher the roller coaster is, the more potential energy it has.

Kinetic energy is the energy an object has because of its motion. The faster the roller coaster moves, the more kinetic energy it has.

At the highest point of the roller coaster, it has the most potential energy. As it goes down the hill and speeds up, its potential energy decreases and its kinetic energy increases. At the lowest point, it has the least potential energy and the most kinetic energy because it is moving the fastest.

Name: ______________________________

Roller Coaster Energy

Look at the picture of the roller coaster below. The roller coaster is at different points labeled with letters (A, B, C, and D).

1. Which letter shows the cart having the maximum kinetic energy? ◯

2. Which letter shows the cart having the maximum potential energy? ◯

3. Which letter shows the cart having the least potential energy? ◯

4. Which letter shows the cart having the least kinetic energy? ◯

5. Which sequence shows an increase in kinetic energy?

   A) B, C, D B) A, B, C
   C) D, C, B D) C, B, A

6. Which correctly shows an increase in potential energy?

   A) From C to D B) From B to C
   C) From A to D D) From D to C

Name: ______________________

Week Seven

Day 4

Vocab Match

Match each word with its definition. Write the correct letter in the space provided.

1. Energy ________
2. Kinetic Energy ________
3. Potential Energy ________
4. Gravity ________
5. Mass ________
6. Speed ________
7. Height ________
8. Friction ________
9. Elastic Energy ________
10. Motion ________

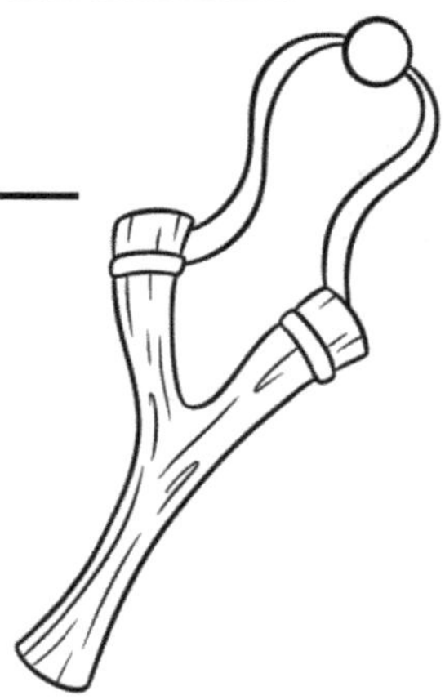

A. Energy that is stored

B. How fast an object is moving

C. Energy stored in objects that are stretched or compressed

D. The force that pulls objects toward the center of the Earth

E. The ability to do work or cause change

F. The action or process of moving or being moved

G. The amount of matter in an object

H. How high an object is, which affects its potential energy

I. The energy an object has because it is moving

J. a force that occurs when two surfaces rub against each other; it acts to slow down or resist motion

Week Seven

Day 5

Use the space below to draw your own roller coaster.
Label key points on your roller coaster with letters A, B, C, and D
Answer the questions based on your roller coaster design.

1. At which point on your roller coaster does the cart have the most potential energy? Why?

2. At which point on your roller coaster does the cart have the most kinetic energy? Why?

3. How would you design certain sections of your roller coaster to maximize speed and acceleration?

(More space for you to record your answers if you need it)

Week Eight

Sound Waves

Did you know? Dolphins and bats use sound waves to navigate and find food! This process, known as echolocation, involves making sounds that bounce off things around them, like an echo. Then they listen to these echoes to figure out where things are, even if it's dark.

Week Eight

Read the passage and analyze the sound wave illustration to complete the tasks.

Understanding Sound Waves

Sound waves are a type of energy that we can hear. They are created when something vibrates, like when you pluck a guitar string or clap your hands. These vibrations make the air around them move, creating waves that travel through the air to our ears.

A sound wave has several important parts:

1. **Crest**: The highest point of a wave.

2. **Trough**: The lowest point of a wave.

3. **Rest Position**: The middle position of the wave when there is no vibration.

4. **Amplitude**: The height of the wave from the resting point to the crest or trough. Higher amplitude means a louder sound.

5. **Wavelength**: The distance from one crest to the next crest. Longer wavelengths make lower-pitched sounds, while shorter wavelengths make higher-pitched sounds.

6. **Frequency**: The number of waves that pass a point in one second. Higher frequency means a higher pitch, and lower frequency means a lower pitch.

When sound waves reach our ears, they make our eardrums vibrate. Our brain then interprets these vibrations as different sounds, like music, voices, or the barking of a dog.

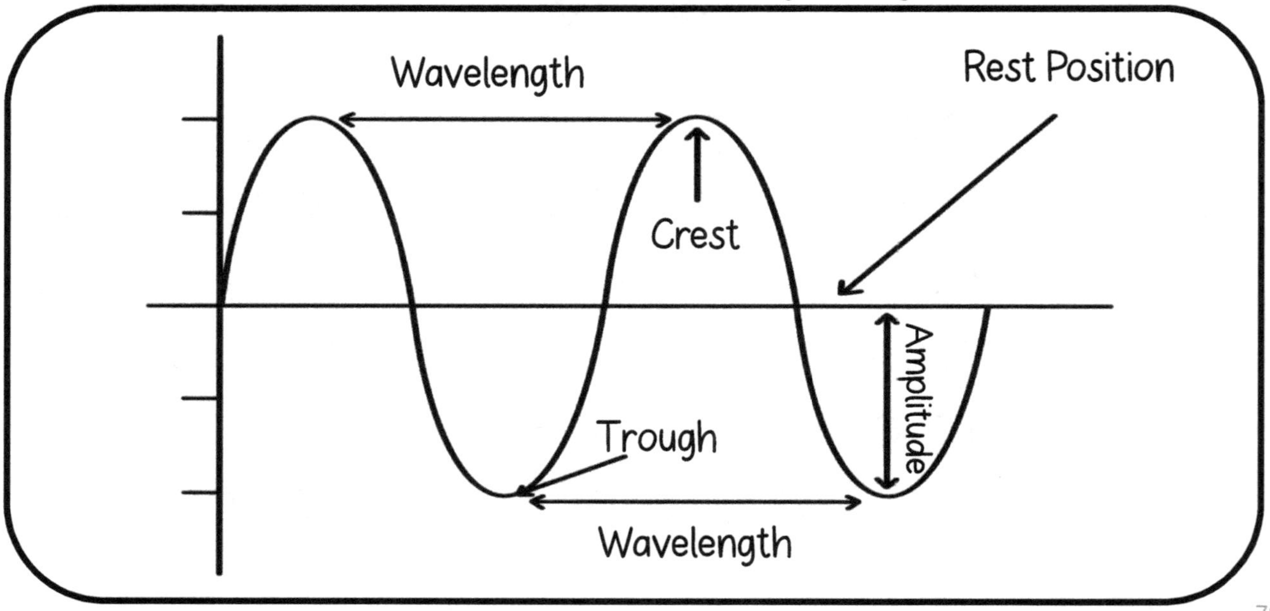

Week Eight

Day 1

Name: ______________________________

Understanding Sound Waves

1. **What creates sound waves?**

   a) Light
   b) Vibrations
   c) Colors
   d) Heat

2. **What is the highest point of a wave called?**

   a) Trough
   b) Frequency
   c) Crest
   d) Amplitude

3. **What does a higher frequency of sound waves mean?**

   a) Higher Pitch
   b) Lower Pitch
   c) Softer sound
   d) No sound

4. **Which part of the sound wave determines how loud the sound is?**

   a) Wavelength
   b) Frequency
   c) Amplitude
   d) Crest

5. **In your own words, explain how sound waves travel from a vibrating object to our ears.**

____

____

____

____

6. **Describe the difference between amplitude and frequency in sound waves. How do they affect what we hear (loudness and pitch)?**

____

____

____

____

Name: ______________________________

Comparing Sound Waves

A **B**

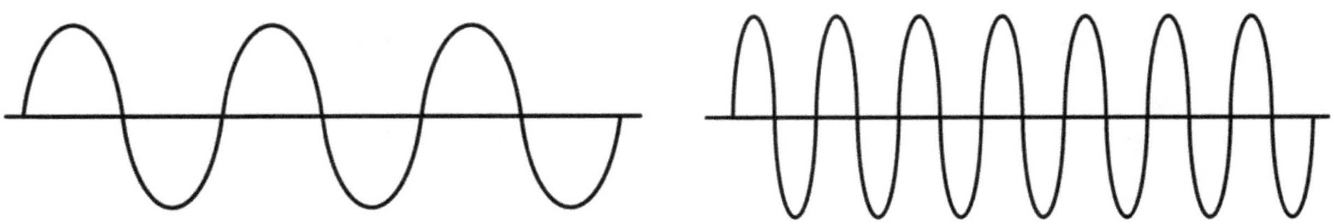

1. Which wave has a higher amplitude?

2. Which wave represents the quieter sound?

3. Which wave has the highest frequency?

4. Which wave has vibrations that are closest together?

5. Which wave is from the highest-pitched sound?

6. Sketch a waveform that would be louder and have a higher pitch than the wave shown below.

Name: ______________________________

Match the Sound Waves

Draw a line from each sound wave to the correct description.

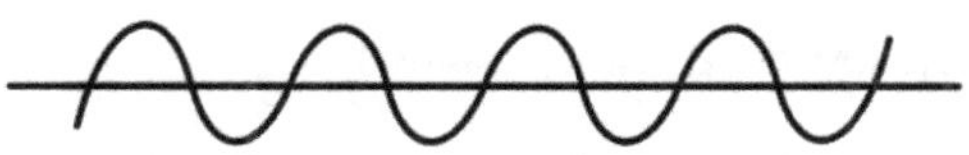

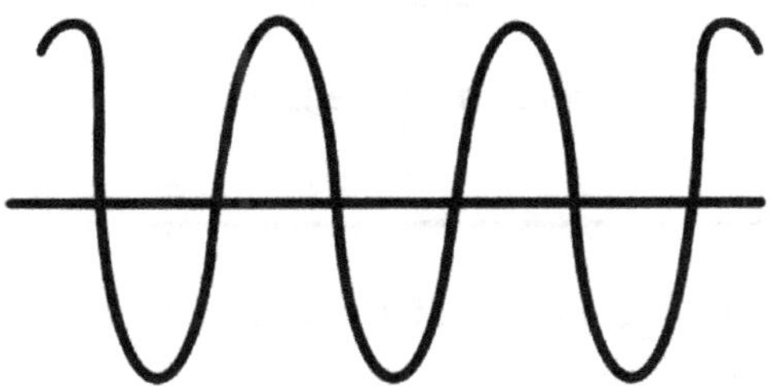

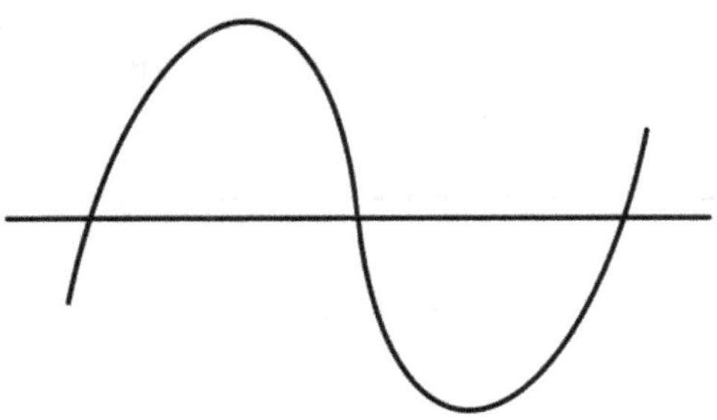

Low pitch, quiet sound

Low pitch, Loud sound

High pitch, quiet sound

High pitch, Loud sound

Name: ____________________

Vocab Match

Match each word with its definition. Write the correct letter in the space provided.

1. Crest ________
2. Trough ________
3. Rest Position ________
4. Amplitude ________
5. Wavelength ________
6. Frequency ________
7. Vibration ________
8. Pitch ________
9. Echo ________
10. Volume ________

A. The number of waves that pass a point in one second

B. The lowest point of a wave

C. How loud or quiet a sound is

D. The height of the wave from the resting point to the crest or trough; determines how loud the sound is

E. The highest point of a wave

F. How high or low a sound is, determined by frequency

G. A back-and-forth movement that creates sound

H. The middle position of the wave when there is no vibration

I. A reflected sound wave that you can hear after the original sound

J. The distance from one crest to the next crest

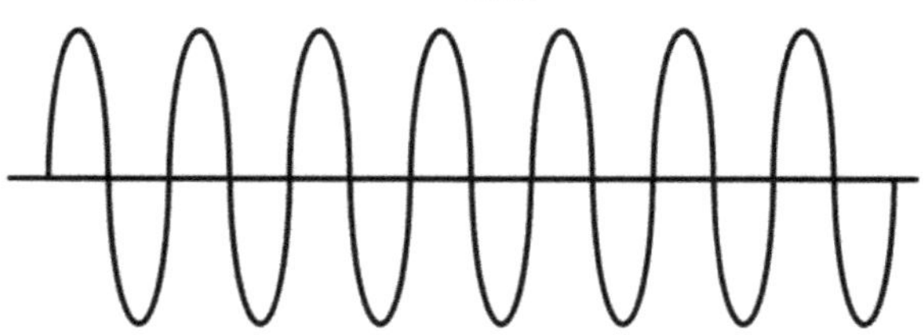

Name: ______________________

Draw and Label the Sound Wave

Read the description below and then draw and label the sound wave that is being described:

Jad excitedly struck a metal tuning fork, and immediately, a loud, high-pitched tone filled the air. The sound was sharp and clear, with the waves moving quickly through the room. Each wave had very high crests and deep troughs, showing the loudness of the sound. The crests were close together, indicating the high pitch of the sound.

Make sure to label the crests, troughs, wavelength, and amplitude.

Week Nine

Magnetism

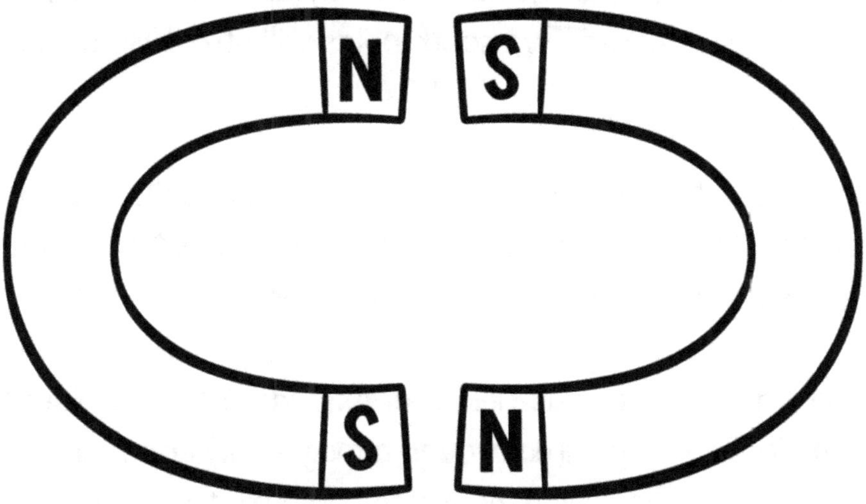

Did you know that about 4000 years ago, according to Greek legend, magnetism was first discovered in ancient Greece? A shepherd named Magnes stumbled upon magnetic rocks (lodestones) when his iron staff got stuck to them. Since this discovery, our use of magnets has evolved in hundreds of different ways.

Week Nine

Read the passage to complete the tasks for the next few days.

The Invisible Force of Magnetism

You can find magnetism at work all around you. Compasses, fans, cars, and computers use magnets to function. Even the world we live on is a giant magnet!

Magnetism is an invisible force that can attract (pull closer) or repel (push away) certain materials. This force is produced by magnets, which come in many shapes and sizes.

A magnet has two ends, called poles: the north pole and the south pole. The north pole of one magnet attracts the south pole of another magnet, but two north poles push away from each other. Two south poles will also push away from each other.

Magnets can attract materials like iron, nickel, and cobalt. These materials are called magnetic materials. Other materials, such as wood, plastic, and paper, are not attracted to magnets and are considered non-magnetic.

The Earth has a giant magnetic field that extends from the planet's interior out into space. A magnetic field is the area around a magnet where magnetic force can be felt. All magnets have a magnetic field, no matter how big they are.

Did you know that Earth's magnetic field causes compass needles to point north? Since opposite poles attract, the south pole of a compass needle is attracted to Earth's magnetic North Pole. Earth's magnetic field also protects us from the Sun's harmful radiation by trapping charged particles.

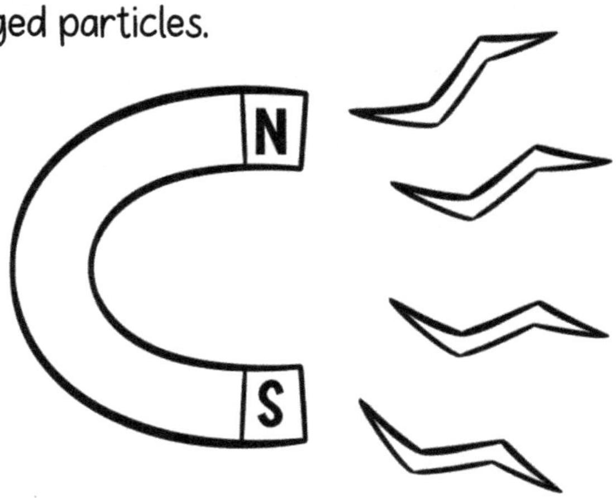

Name: ______________________

Week Nine

Day 1

After reading the passage "The Invisible Force of Magnetism", answer the questions.

1. **What are the two ends of a magnet called?**
   - a) Tips
   - b) Edges
   - c) Poles
   - d) Corners

2. **Which materials are attracted to magnets?**
   - a) Wood, plastic, and paper
   - b) Iron, nickel, and cobalt
   - c) Glass, aluminum, and rubber
   - d) Copper, gold, and silver

3. **What happens when the north pole of one magnet is brought near the south pole of another magnet?**
   - a) Nothing happens
   - b) They neutralize each other
   - c) They repel each other
   - d) They attract each other

4. **What causes compass needles to point north?**
   - a) The Earth's magnetic field
   - b) The Earth's rotation
   - c) Gravity
   - d) The sun's radiation

5. **According to the passage, what role do magnets play in everyday objects? Provide at least three examples.**

MAGNETISM

Week Nine

Day 2

Name: ______________________________

Magnetic or Non-Magnetic

Determine if each object is magnetic or non-magnetic. Color the thumbs-up icon if it is magnetic, and color the thumbs-down icon if it is non-magnetic.

1.

Plastic Spoon

2.

Iron Nail

3.

Steel Paperclip

4.

Rubber Eraser

5.

Staples

6.

Glass Marble

7.

Wooden Pencil

8.

Scissors

9.

Aluminum Foil

10.

Steel Push Pin

MAGNETISM

Name: ____________________

Week Nine

Day 3

Attract or Repel

When the north pole of a magnet is near the south pole of another magnet, they attract (pull closer). When like poles of two magnets are near each other, such as north to north or south to south, they will repel (pull away) each other.

Determine whether each pair of magnets will attract or repel based on the poles shown.

1.

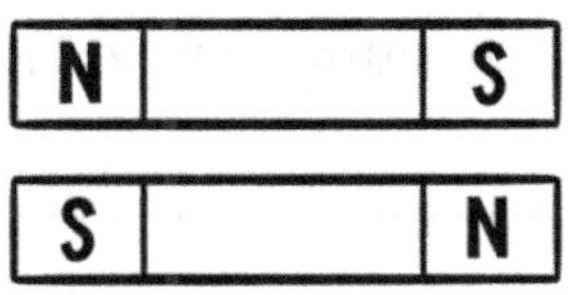

____

2.

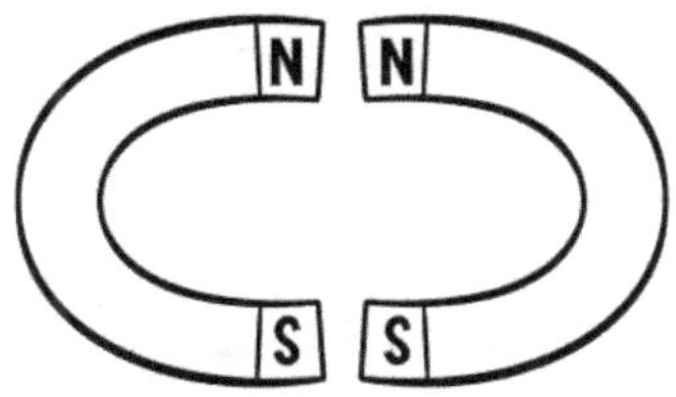

____

3.

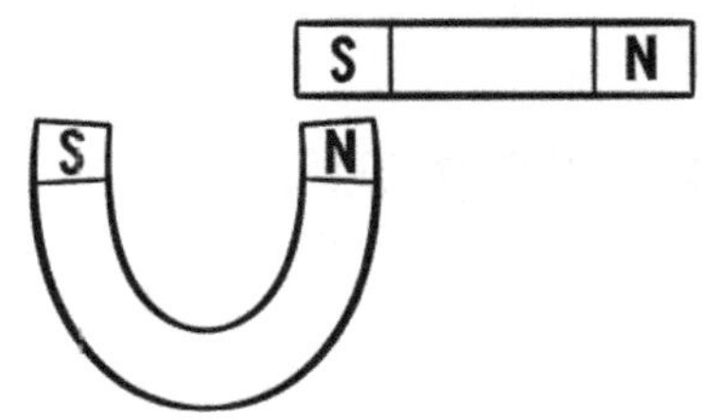

____

4.

____

5.

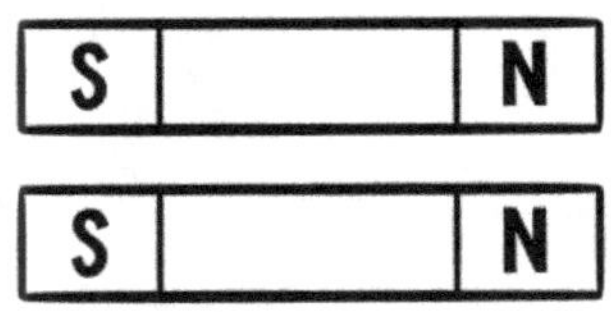

____

6.

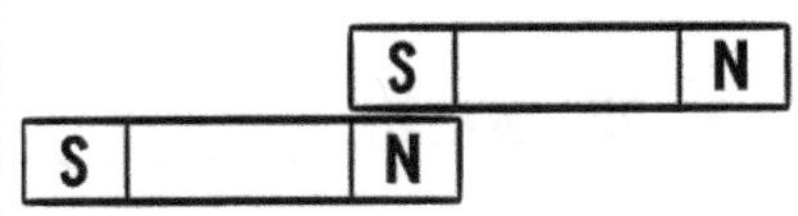

____

7.

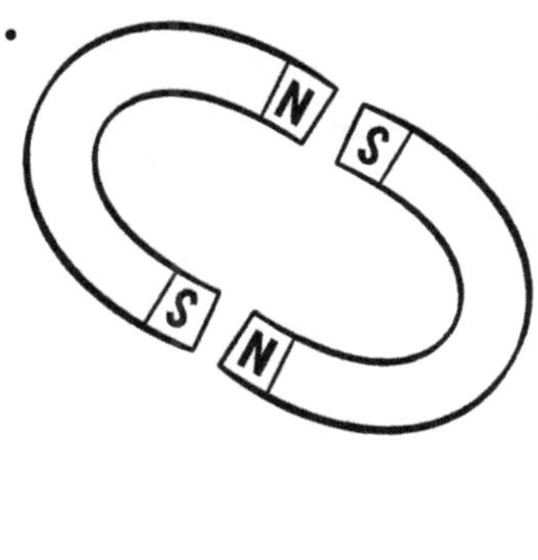

____

8.

____

9.

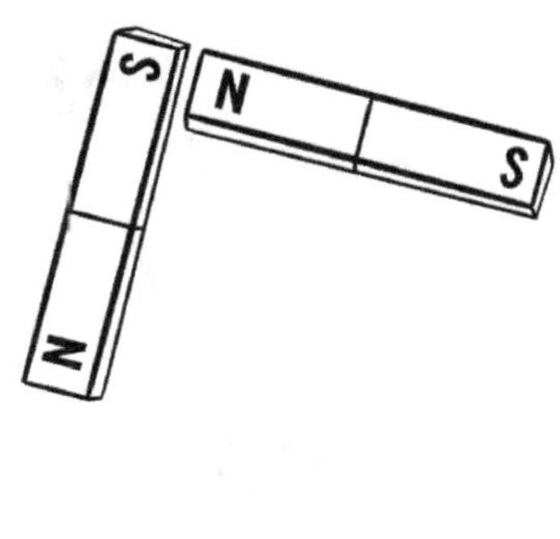

____

Name: ______________________

Week Nine

Day 4

Vocab Match

Match each word with its definition. Write the correct letter in the space provided.

1. Magnet ________
2. Magnetic Field ________
3. Magnetism ________
4. Attract ________
5. Repel ________
6. Poles ________
7. Compass ________
8. Iron ________
9. Electromagnet ________

A. The invisible force that causes magnets to attract or repel objects

B. To push away

C. An object that can attract iron and some other materials

D. A tool that uses a magnetized needle to show direction by pointing to the Earth's magnetic north.

E. A type of magnet created by electric current flowing through wire coiled around a metal core

F. The area around a magnet where magnetic forces can be felt.

G. The ends of a magnet where the magnetic force is the strongest

H. To pull towards

I. A type of metal that is strongly attracted to magnets

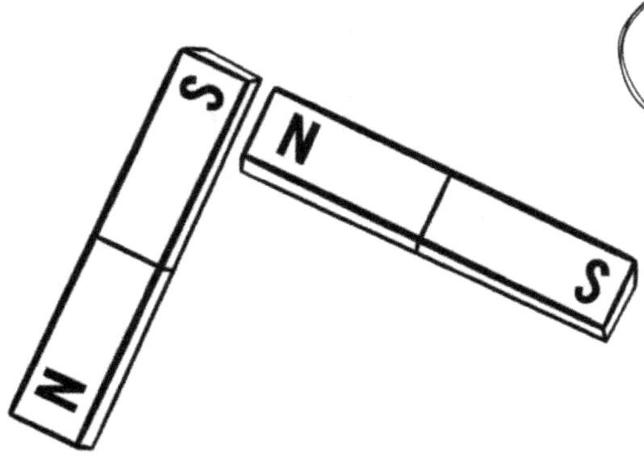

Week Nine

Day 5

Describe a day without magnets. What everyday items or activities would be affected if magnets suddenly stopped working? How would life be different?

Exploring Life Science

Week One

Food Chains and Food Webs

Every creature depends on each other for food. It's like a wild game of Jenga. Once one piece is gone, the rest start to wobble!

Week One

Day 1

Name: ______________________________

Producers, Consumers, and Decomposers

Read the passage, then label the organisms.

All living things have important roles in an ecosystem and rely on each other as part of a food chain. A food chain shows how living things get their energy and nutrients to survive. Producers, consumers, and decomposers are the main parts of a food chain. Each group has a special role based on what they eat and how they help the ecosystem.

Plants are called **producers** because they make their own food using energy from sunlight in a process called photosynthesis. **Consumers**, such as animals, cannot make their own food and must eat other living things for energy. Animals that eat only plants are called **herbivores**. **Carnivores** eat other animals, and **omnivores** eat both plants and animals. **Decomposers** break down dead plants and animals and return nutrients to the soil.

All living things depend on each other to keep ecosystems healthy and balanced!

Identify and label each organism. Use the sentence under each picture to help you choose the correct label.

WORD BANK

Carnivore

Producer

Herbivore

Omnivore

Decomposer

1.

Rabbits eat grass and hay.

____

2.

Earthworms break down dead plants and animals

____

3.

Brown bears eat plants, berries, fish and insects

____

4.

Snakes eat other animals and sometimes other snakes.

____

5.

Lions mainly eat meat.

____

6.

Sunflowers make their food through photosynthesis.

____

7.

Giraffes eat leaves, buds, and twigs from trees.

____

8.

Mushrooms break down dead leaves, wood, and plant debris.

____

Week One

Day 2

Name: ______________________________

Analyze the food chain and food web, then answer the following questions.

Food Chain

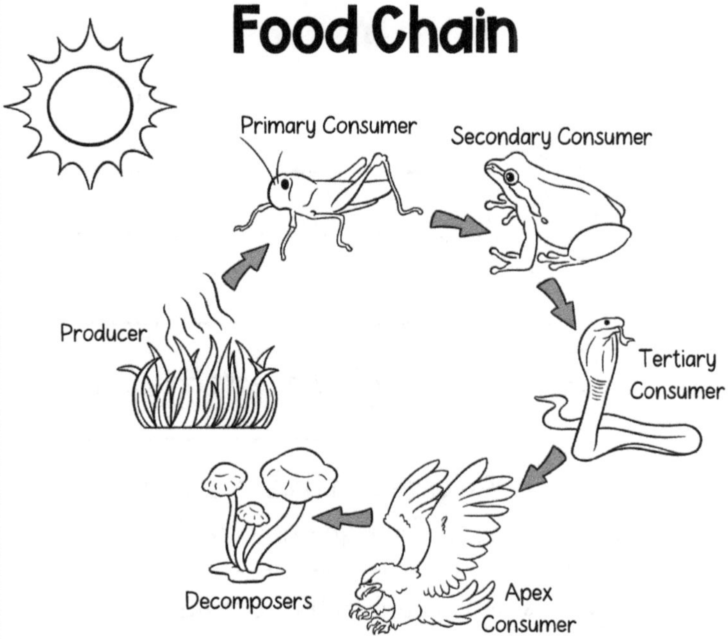

A food chain only follows one path as animals find food.

Food Web

A food web consists of many food chains.

1. What is the primary source of energy in a food chain or food web?

a) Water
b) Sunlight
c) Air
d) Rocks

2. What is a consumer in a food chain or food web?

a) An organism that eats other organisms for food.
b) An organism that uses sunlight to make its own food.
c) An organism that breaks down dead organisms.
d) An organism that produces oxygen for the ecosystem.

3. Which of the following best describes a food chain?

a) A group of animals living together in a specific habitat.
b) A process by which plants make their own food using sunlight
c) A sequence of organisms where each one depends on the next as a source of food.
d) A protective covering that helps some animals blend into their environment.

4. What do arrows represent in a food chain or food web?

a) The competition between different species.
b) The transfer of energy from one organism to another.
c) The growth and development of plants and animals.
d) The process of photosynthesis.

Week One

Day 3

Name: ______________________

Vocab Match

Match each word with its definition. Write the correct letter in the space provided.

1. Producer ____________
2. Omnivore ____________
3. Herbivore ____________
4. Carnivore ____________
5. Decomposers ____________

A. A consumer that eats only other animals

B. A consumer that eats both plants and animals

C. A consumer that eats only plants

D. An organism that makes its own food through photosynthesis

E. An organism that breaks down dead plants and animals and returns nutrients to the soil

6. Complete the chart below.

Herbivores	Carnivores	Omnivores	Decomposers
List 2 examples ☆ ☆	List 2 examples ☆ ☆	List 2 examples ☆ ☆	List 2 examples ☆ ☆
Illustrate the foods they eat	Illustrate the foods they eat	Illustrate the foods they eat	Illustrate the foods they eat

Week One

Day 4

Name: ______________________________

Exploring a Food Web

Use the food web below to answer the questions.

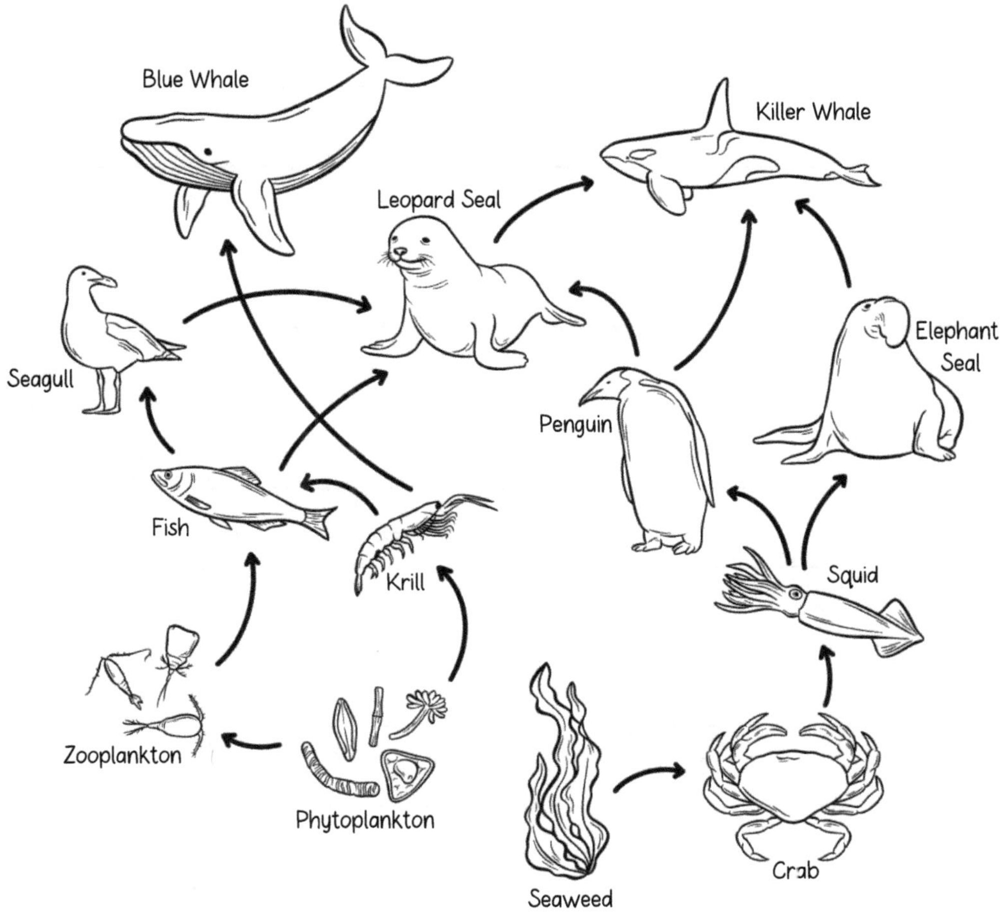

1. What are the producers in this food web?

2. What are the top consumers in this food web? Explain how you know.

3. How many different food chains can you trace from the producers to the top consumers? Count each complete path.

Name: ______________________________

Week One

Day 5

Why do you think a disruption in one part of a food chain or food web can affect other organisms? Explain your thinking.

Week Two

Biomes

In the Amazon rainforest, there are more species of plants and animals than anywhere else on Earth. But, did you know that scientists are still discovering new species every day?

Name: ______________________

Week Two

Day 1

The Fascinating World of Biomes

Biomes are large regions on Earth where the climate, plants, and animals share similar characteristics. Each biome has its own temperature, weather, and types of living things that can survive there.

- **Desert** biomes are hot and very dry with very little rainfall; only certain plants, like cacti, can survive the harsh heat.
- **Tropical rainforest** biomes have lots of rainfall year-round and tall trees that form a thick canopy. Many animals, such as monkeys and colorful birds, live high in the trees.
- **Grassland** biomes are large, open areas covered in grasses. They get some rain but not enough to support many trees.
- **Tundra** is the coldest biome, with frozen ground called permafrost and very few plants.
- **Marine** biomes include Earth's oceans. They cover about 70% of Earth's surface and are home to many ocean animals, such as blue whales and sharks.
- **Freshwater** biomes include rivers, lakes, and ponds; their water is not salty.

Each biome provides a habitat for plants and animals that have adapted to live in its special conditions.

Identify and label each biome.

1. The coldest biome

____

2. Includes many tall trees and high rainfall

____

3. The biggest biome covering about 70% of the Earth

____

4. The driest biome on Earth

____

5. Includes rivers, lakes, and ponds

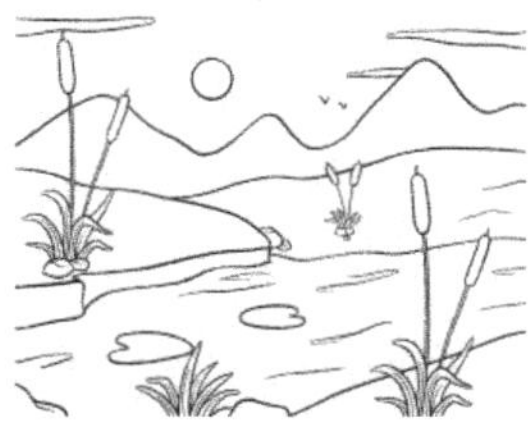

____

6. Large and very flat areas covered in grass

____

Week Two

Day 2

Name: ____________________

Use what you learned from yesterday's reading and additional research to answer the questions below. Some answers may require you to look up information about different biomes.

1. **Which biome is known for its tall trees that lose their leaves in the fall?**
   a) Rainforest
   b) Desert
   c) Tundra
   d) Deciduous Forest

2. **In which biome can you find blue whales and sharks?**
   a) Tundra
   b) Marine
   c) Freshwater
   d) Desert

3. **Which biome has lots of coniferous trees like pine and spruce?**
   a) Taiga
   b) Tundra
   c) Desert
   d) Grassland

4. **Which biome is characterized by its vast grassy plains, with grazing animals like bison and zebras?**
   a) Rainforest
   b) Desert
   c) Grassland
   d) Tundra

5. Describe the freshwater biome. Include information about the types of plants and animals you might find, and then illustrate it.

Describe it.	Draw it.
____________________ ____________________ ____________________ ____________________ ____________________ ____________________ ____________________ ____________________ ____________________ ____________________	

Name: ______________________

Week Two

Day 3

Match each word with its definition. Write the correct letter in the space provided.

1. Desert ________
2. Rainforest ________
3. Tundra ________
4. Grassland ________
5. Deciduous Forest ________

(A) A cold and icy biome found in the far north that has frozen ground called permafrost

(B) A dry and sandy or rocky place where it doesn't rain much

(C) Goes through the four seasons; trees lose their leaves at the end of each growing season

(D) Wide open spaces with lots of grass and few trees

(E) A dense and lush forest found in warm regions with a lot of rainfall

6. Compare and contrast tropical rainforests and temperate rainforests by listing their differences and similarities.

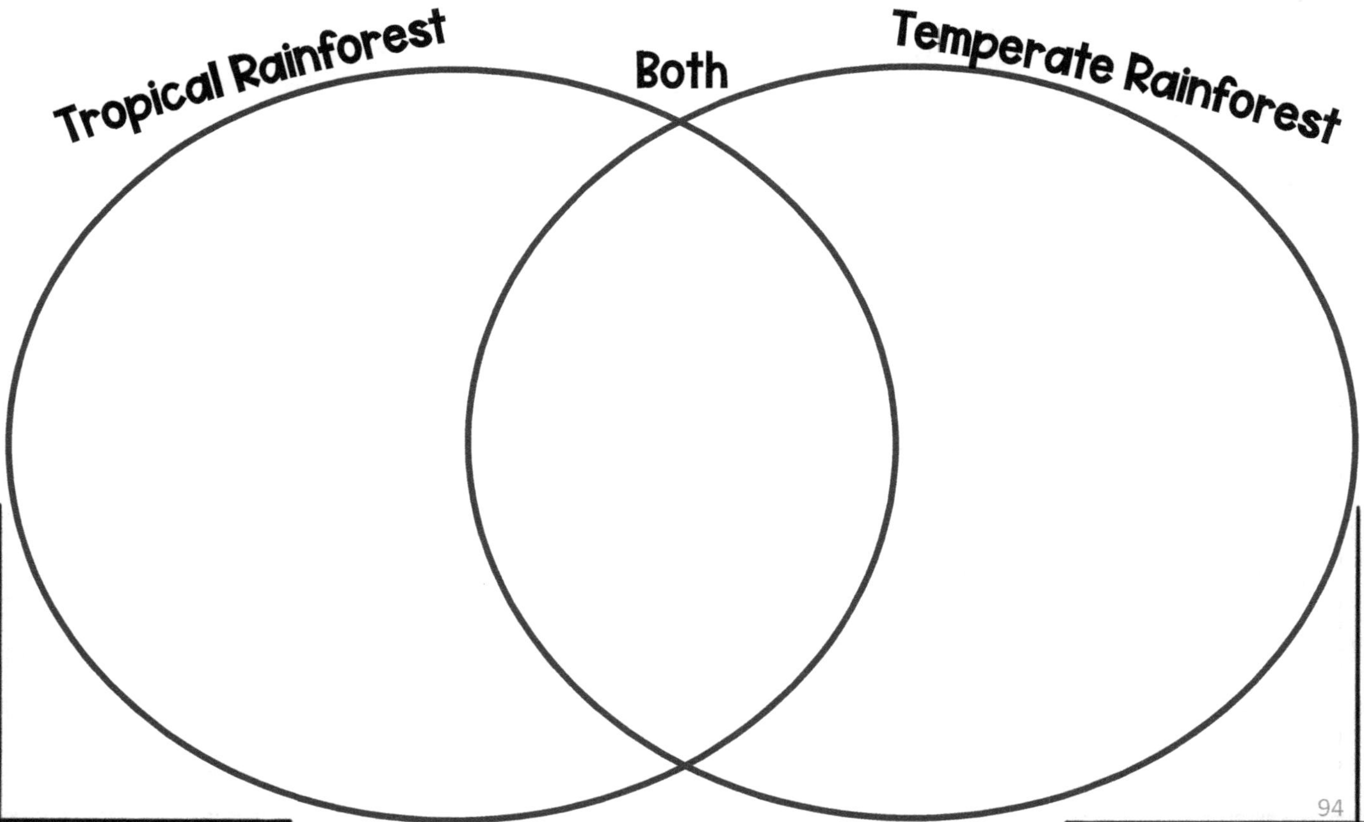

Name: ______________________________

Guided Biome Map Activity

Scan the QR code using a tablet or phone. Look closely at the biome map and notice how each region is colored. Color the same biome areas on your map.

SCAN ME

If QR access is unavailable, display a world biome map or provide a printed reference. Focus student attention on patterns such as proximity to the equator, poles, and oceans.

1. Color in the locations of the tropical rainforests green.
2. Color in the locations of the deserts brown.
3. Color in the locations of the tundra orange.
4. Color in the locations of the deciduous forests yellow.
5. Color in the locations of the grasslands red.
6. Color in the locations of the taiga purple.

What is one new thing you learned about where biomes are located on Earth today?

____

____

____

____

Name: ______________________________

Week Two

Day 5

If you were a scientist studying the desert biome, what questions would you want to investigate? What would you like to learn about the plants and animals that live in the desert? What questions do you have about the adaptations that help them survive?

Week Three

Animal Adaptations

Animals are constantly adapting to changes in their environment.

Week Three

Name: ______________________

Instinctive and Learned Behavior

Animal behaviors can be divided into two main categories: **instinctive** and **learned**. Instinctive behaviors are actions animals are born with and perform automatically, without needing to be taught. For example, when a baby bird hatches from its egg, it instinctively knows how to open its mouth for food and chirp for its parents' attention.

Migration and hibernation are also instinctive behaviors, meaning animals are born knowing how to do them. Some animals travel long distances to find food or warmer weather, while others sleep through cold winter months.

Learned behaviors, on the other hand, are actions animals develop over time through experience and observation. These behaviors are not automatic and require practice. A dog learning to sit or fetch in response to commands from its owner is an example of a learned behavior. Both instinctive and learned behaviors help animals survive and succeed in their environments.

Identify and label each animal behavior as instinctive or learned.

1. 

Baby sea turtles move to the ocean when they hatch.

____

2. 

Birds migrating to search for food

____

3. 

A dog hears a bell and comes for food.

____

4. 

A spider builds a web.

____

5. 

Dolphins performing tricks

____

6. 

A parrot saying "hello".

____

Name: ______________________________

Read the Passage and answer the questions.

Elephants

Did you know that elephants are the biggest land animals in the world? They can grow up to eleven feet tall and weigh as much as 13,000 pounds. Yikes! Elephants are known for their long, muscular trunks. Their trunks help them in many ways. Elephants use their trunks to pick up food, move heavy objects, and suck up water to drink. They can even spray water over their bodies to cool off. Elephants also throw dirt over their backs to protect their skin from the sun and insects.

Elephants have one large tooth on each side of their upper jaw that forms a tusk. Each tusk can measure up to seven feet in length. Elephants use their tusks for digging, fighting, and feeding. For example, they scrape bark off trees using their tusks. Their tusks can also support their trunks when resting.

Have you ever noticed how big elephants' ears are? Everything has a purpose in nature, including elephant ears! When elephants get too hot, they flap their ears to create air currents and help release extra body heat. Flapping your ears sounds like a fun way to cool off, don't you think?

Unfortunately, because of their size and their valuable ivory tusks, elephants are often hunted. This has caused their population to decrease, and they need protection to survive in the future.

1. What physical feature allows the elephant to dig, fight, and feed?

____

2. What physical feature allows the elephant to pick up food?

____

3. Why do elephants throw dirt over their backs?

____

____

4. What does it mean when elephants flap their big ears?

____

____

Week Three

Day 3

Name: ______________________________

Physical and Behavioral Adaptations

Animals have adaptations that help them survive in their environments. These adaptations can be grouped into two main types: physical adaptations and behavioral adaptations.

Physical adaptations are body parts or features that animals are born with. These features help animals survive in their habitats. For example, a giraffe has a long neck to reach high leaves for food.
Camouflage is a physical adaptation that helps an animal blend into its surroundings by matching the colors or patterns around it. Mimicry is another physical adaptation where an animal looks like another organism to stay safe from predators.

Behavioral adaptations are actions animals take to survive n their environment. These behaviors describe what an animal does, not what it looks like. For example, birds fly south for the winter to find warmer weather and more food.

Classify whether each fact describes a physical or behavioral adaptation. Write physical or behavioral under each fact.

1.

Polar bears have thick fur to keep them warm.

____

2.

A bird builds a nest to lay eggs.

____

3.

Eagles have sharp claws to help them grasp prey

____

4.

Butterflies migrate to find food

____

5.

Fish travel in groups for protection

____

6.

Penguins have webbed feet for efficient swimming in water.

____

Week Three

Day 4

Name: ______________________

Vocabulary Match

Match each word with its definition. Write the correct letter in the space provided.

1. Physical Adaptation _______
2. Behavioral Adaptation _______
3. Camouflage _______
4. Mimicry _______
5. Instinctive Behavior _______
6. Learned Behavior _______
7. Hibernation _______
8. Migration _______
9. Predator _______
10. Prey _______

A. Behaviors animals are born with and perform naturally

B. When animals travel long distances to find warmer weather, food, or safe places to raise young

C. Something an animal does to help it survive in its environment

D. An animal that is hunted and eaten by another animal for food

E. Looking or sounding like another living organism

F. Behaviors learned through experience and observation

G. A visual disguise where an animal's color or pattern helps it blend into its surroundings

H. A body feature or structure that helps an animal survive in its habitat

I. An animal that hunts and eats other animals for food

J. When an animal sleeps or becomes inactive during cold winter months

Week Three

Day 5

Name: ______________________________

Animals have physical and behavioral adaptations that help them survive and find food. Choose one animal and describe one physical adaptation and one behavioral adaptation it uses to get food. Explain how each adaptation helps the animal survive in its environment.

Week Four

Animal Classification

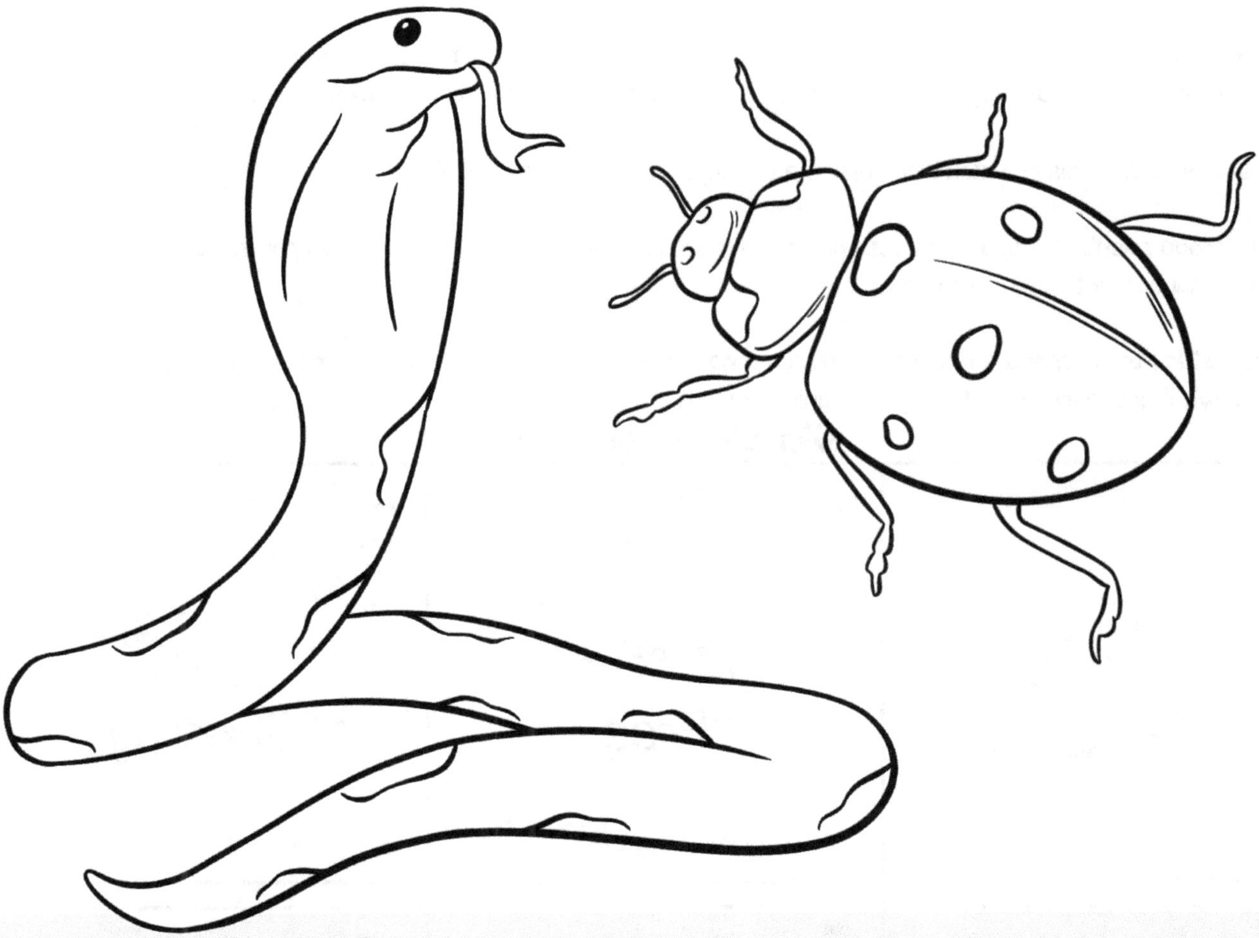

Classifying animals helps us understand the diversity of life on Earth.

Week Four

Day 1

Name: ______________________________

How Scientists Group Animals

There are many different types of animals on Earth, and scientists classify them into groups based on their characteristics. Some animals have fur or hair, give birth to live babies, and feed their young with milk. These animals are called **mammals.**

Animals with feathers, wings, beaks, and two legs are classified as **birds**. Most birds can fly, but some, such as penguins and ostriches, cannot. All birds lay eggs. and their eggs protect the growing baby bird inside until it is ready to hatch.

Animals covered in scales and with dry skin are known as **reptiles.** They usually lay eggs on land and can live in many different environments.

Amphibians spend part of their lives in water and part on land. They are born with gills to breathe underwater, but later they develop lungs as they grow into adults. Their smooth, moist skin helps them survive in water.

Fish live their entire lives in water. They breathe through gills and swim using fins.

Another group scientists study is **insects**. Insects have six legs and a body with three main parts: the head, thorax, and abdomen. Many insects also have wings.

Scientists study these characteristics to understand how different animals survive in their environments. Now it's your turn. Use what you learned to classify each animal below.

Classify each animal.

1. 

Snake

____

2. 

Frog

____

3. 

Stingray

____

4. 

Giraffe

____

5. 

Penguin

____

6. 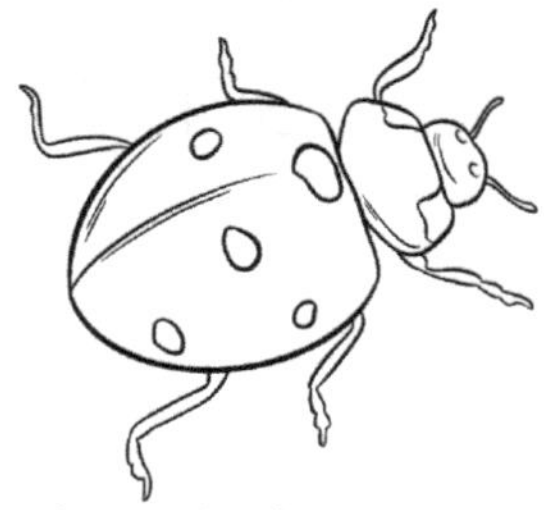

Ladybug

____

Name: ______________________________

Week Four

Day 2

Warm-Blooded and Cold-Blooded Animals

Warm-blooded and cold-blooded are terms used to describe how animals regulate their body temperature.

Warm-blooded animals maintain a constant body temperature regardless of the temperature outside. Mammals and birds are warm-blooded animals.

Cold-blooded animals' body temperature changes based on the temperature of their environment. Amphibians, reptiles, insects, and fish are cold-blooded animals.

List 5 warm-blooded animals and 5 cold-blooded animals in the chart below. Then write what each animal is classified as (mammal, bird, reptile, amphibian, fish, or insect).

Warm-Blooded Animals	Cold-Blooded Animals
1.	1.
2.	2.
3.	3.
4.	4.
5.	5.

Week Four

Day 3

Name: ______________________

Vertebrate and Invertebrate Animals

Vertebrates are animals that have a backbone or spine. Mammals, birds, reptiles, amphibians, and fish are vertebrates.

Invertebrates are animals that DO NOT have a backbone or spine. Insects, worms, and jellyfish are invertebrates.

Decide whether each animal is a vertebrate or an invertebrate. Write your answer under each picture.

1.

Frog

____

2.

Bird

____

3.

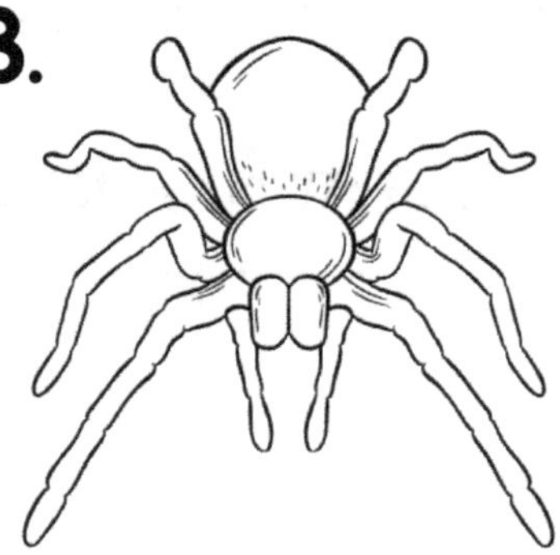

Spider

____

4.

Horse

____

5.

Earthworm

____

6.

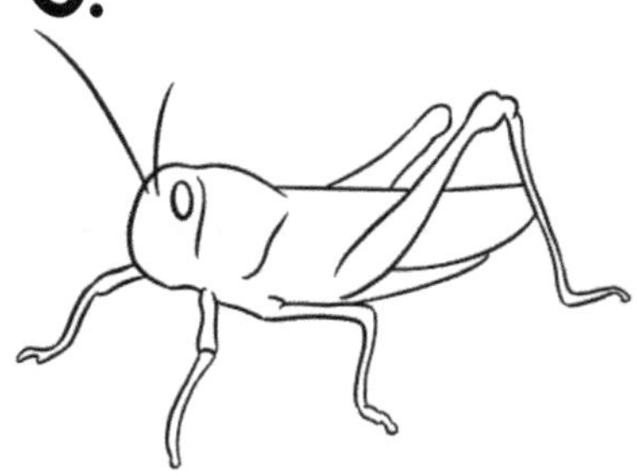

Grasshopper

____

Week Four

Day 4

Name: ______________________

Vocabulary Match

Match each word with its definition. Write the correct letter in the space provided.

1. Vertebrate ________
2. Invertebrate ________
3. Warm-blooded ________
4. Cold-blooded ________
5. Insects ________
6. Mammals ________
7. Amphibians ________
8. Reptiles ________
9. Fish ________
10. Birds ________

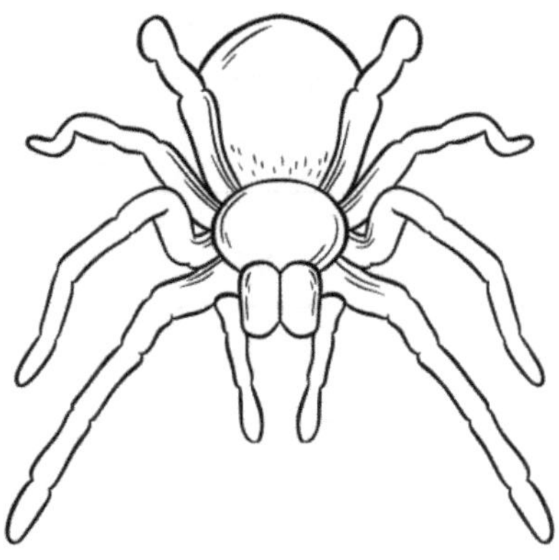

A. An animal's body temperature changes depending on the temperature of its environment

B. Warm-blooded vertebrate animals that have fur or hair, feed their young with milk, and breathe air with lungs

C. Warm-blooded animals with feathers, wings, and a beak that lay eggs and usually fly

D. Animals that maintain a constant body temperature regardless of the environment

E. Cold-blooded vertebrate animals with scaly skin that usually lay eggs on land

F. An animal that has a spine or backbone

G. Invertebrate animals with three body parts (head, thorax, and abdomen), six legs, and often wings

H. Cold-blooded animals that live in water and use gills to breathe and fins to swim

I. An animal that does not have a spine or backbone

J. Cold-blooded vertebrate animals that live part of their life in water and part on land

Name: ______________________________

Week Four

Day 5

Imagine you discover a new animal in the jungle. Describe its physical features and one behavior that helps it survive. Then classify the animal as a mammal, bird, reptile, amphibian, fish, or insect. Explain whether it is warm-blooded or cold-blooded and whether it is a vertebrate or invertebrate. Use what you learned this week to explain your thinking.

Name: ______________________________

Week Four

Day 5

Draw the unique animal you described below.

Week Five

Plant and Animal Cells

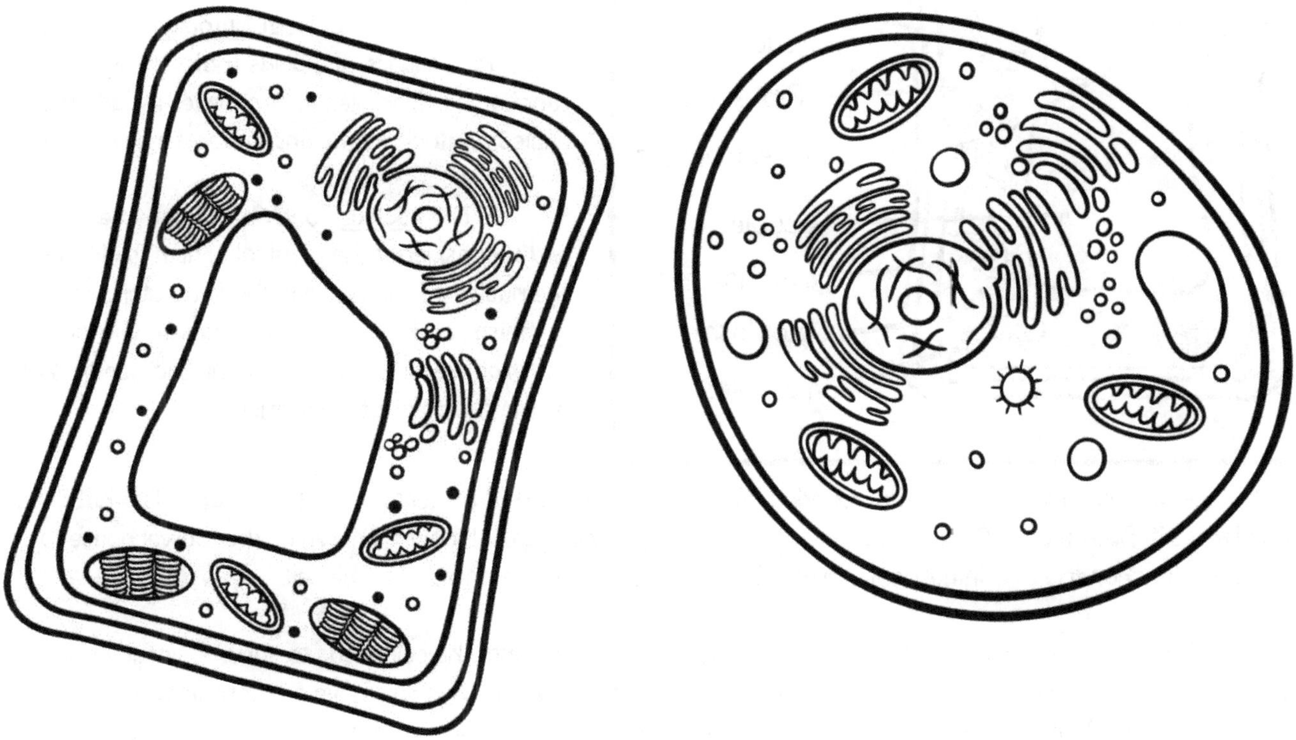

Plant cells have tiny green structures called chloroplasts that act like solar panels, turning sunlight into energy through photosynthesis.

Week Five

Examine the plant cell and read the passage. Use the passage to complete the tasks for the next few days.

Plant and Animal Cells

Plant Cell

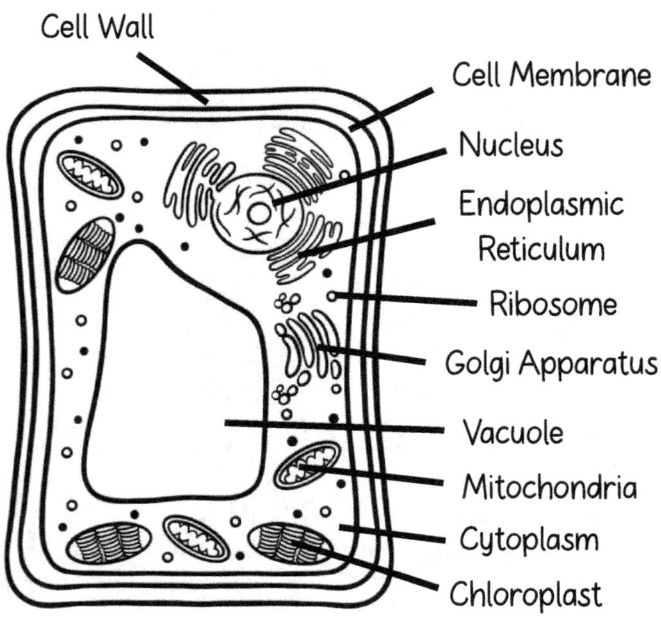

Cells are the basic structures of all plants and animals. Did you know that you are made up of cells? Cells are very tiny, so you need a microscope to see them. A scientist named Robert Hooke first discovered cells in 1665.

Both plant and animal cells are made up of organelles that perform specific jobs inside the cell. Organelles are protected and held inside a gel-like substance called cytoplasm. Cytoplasm contains fluids, fats, sugars, and acids that help keep the cell functioning. Examples of organelles include the nucleus, mitochondria, and endoplasmic reticulum.

The nucleus controls the activities of the cell and is considered the "control center." DNA is found in the nucleus. It determines the characteristics of an organism, such as what the organism will look like. For example, DNA can determine the size of a plant or the eye color of an animal.

Plant and animal cells are surrounded by a cell membrane that protects the cell and controls what enters and leaves it. Both types of cells get their energy from mitochondria, which are known as the "powerhouse" of the cell. Mitochondria take in nutrients, break them down, and release energy for the cell.

Vacuoles are storage tanks that hold water, food, and nutrients the cell needs to survive. Vacuoles are larger in plant cells than in animal cells because they store more water. Cells also have a transportation network called the endoplasmic reticulum that moves proteins and lipids to the Golgi apparatus. The Golgi apparatus is known as the "post office" because it sorts and processes proteins. Ribosomes are tiny structures found in the cell that make proteins. Proteins help build new structures, repair damage, and direct chemical reactions.

Although plant and animal cells share many similarities, there are differences too. Plant cells have a cell wall, but animal cells do not. The cell wall is an extra layer that helps maintain the cell's shape. Plant cells also contain chloroplasts, which animal cells do not have. Chloroplasts help plants use energy from sunlight to make food.

Plant and animal cells are different in shape and size too. Plant cells are usually larger and rectangular, while animal cells are round or irregular in shape.

Understanding the parts of cells helps scientists learn how living things grow, survive, and stay healthy.

Week Five

Day 1

Name: ____________________

Label the Animal Cell.

Refer back to the reading on plant and animal cells.
Use the descriptions of each organelle and the word bank to correctly label the animal cell diagram.

Word Bank

Endoplasmic Reticulum	Cytoplasm	Golgi Apparatus	Nucleus
Cell Membrane	Ribosomes	Vacuole	Mitochondria

Animal cells do not have a cell wall or chloroplasts.

PLANT AND ANIMAL CELLS

Week Five

Day 2

Name: ______________________________

After reading about plant and animal cells, answer the following questions.

1. **Which of the following is a characteristic of plant cells but not animal cells?**

   a) Nucleus
   b) Mitochondria
   c) Vacuole
   d) Cell Wall

2. **Which organelle is responsible for providing energy to the cell?**

   a) Nucleus
   b) Mitochondria
   c) Ribosome
   d) Golgi Apparatus

3. **Which organelle is often referred to as the "control center" of the cell?**

   a) Golgi Apparatus
   b) Endoplasmic Reticulum
   c) Nucleus
   d) Vacuole

4. **What is the function of the vacuole in plant cells?**

   a) It stores water and nutrients for the cell.
   b) It provides the main structure and protection for the cell.
   c) It helps with the production of energy in the cell.
   d) It controls the movement of substances in and out of the cell.

5. **What is the transportation network of the cell called?**

   ____

6. **What is considered the "post office" of the cell?**

   ____

7. **What organelle makes proteins?**

   ____

8. **Why do cells need proteins? Explain using information from the text.**

   ____

   ____

Name: ______________________

Week Five

Day 3

Organelle Function

Write the function for each organelle. The first one is done for you.

Organelle	Function
Mitochondria	Breaks down food to release energy for the cell
Cell Membrane	
Vacuole	
Chloroplast	
Golgi Apparatus	
Ribosomes	
Endoplasmic Reticulum	
Nucleus	
Cytoplasm	
Cell Wall	

Week Five

Day 4

Name: ______________________

Plant Cell, Animal Cell, or Both?

Classify each statement as belonging to a **plant cell**, an **animal cell**, or **both**.

1.

Has a nucleus

____

2.

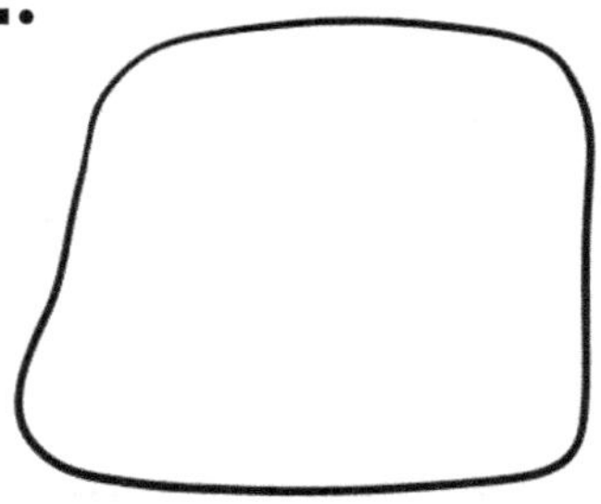

Has a large vacuole

____

3.

Contains chloroplasts

____

4.

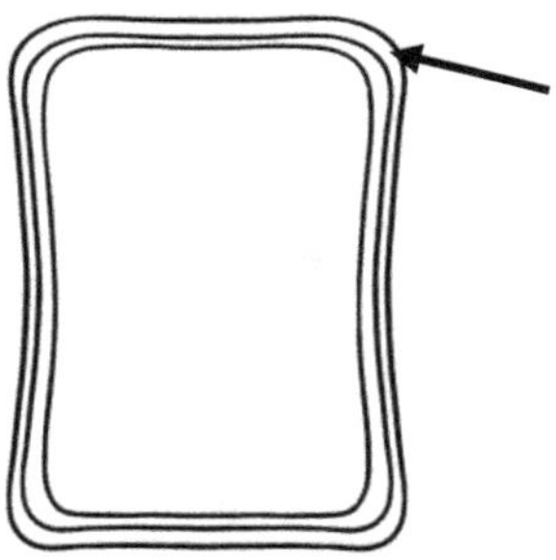

Has a cell wall

____

5.

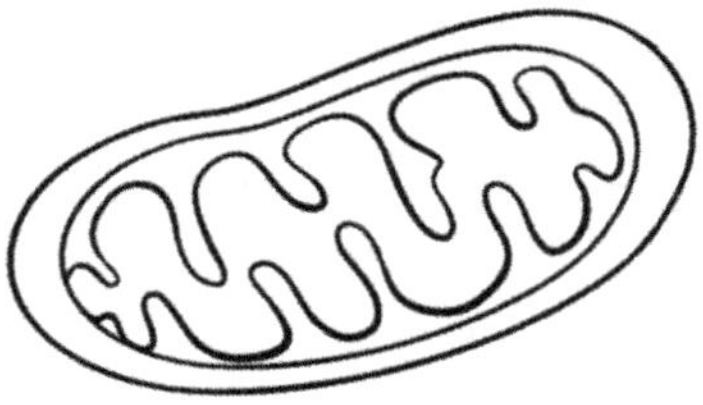

Contains mitochondria

____

6.

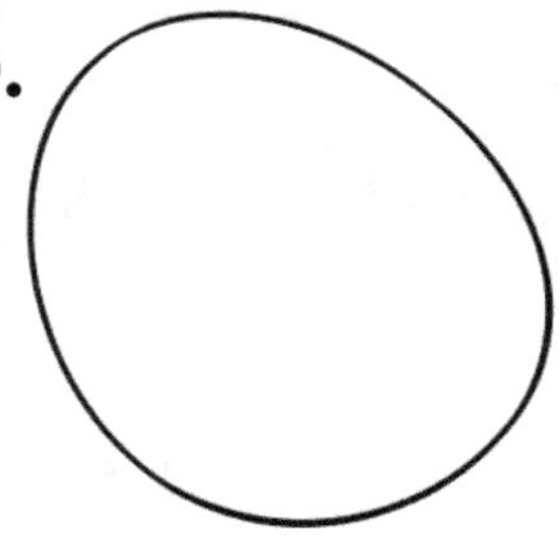

Its shape is round or irregular

____

Name: ______________________________

Imagine you are a plant cell. Describe the importance of chloroplasts and how they help you survive and grow. How would your cell be different if it did not have chloroplasts?

Week Six

Human Body Systems

Your blood vessels are like highways, transporting oxygen, nutrients, and waste products throughout your body. Did you know that if you stretched out all the blood vessels in your body, they would circle the Earth more than twice?

Name: ______________________________

The Digestive System

Read and gather clues from the passage to label the Digestive System.

Did you know that everything you eat is broken down into substances your organs and cells can use? That is the job of your digestive system — breaking down food into nutrients that can be absorbed by the body. Depending on what you eat, it can take about 24 hours for food to move through the digestive system, which is around 30 feet long!

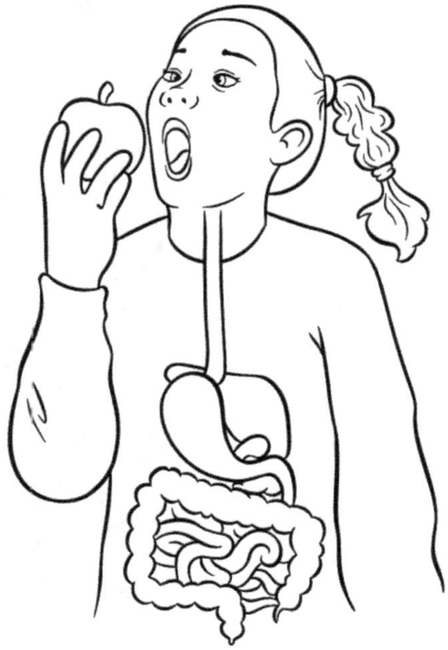

The digestive system begins with the mouth, where food is chewed and mixed with saliva, making it easier to swallow. From there, food travels down the esophagus and into the stomach. Did you know that your stomach is about the size of a tennis ball when it is empty and can stretch to the size of a football when it is filled with food? The stomach releases digestive juices and acid that break down food even more and kill harmful bacteria so that you don't get sick.

Next, the partially digested food moves into the small intestine, where most nutrient absorption occurs. The small intestine is not really that small! If you stretched out an adult's small intestine, it would be about 22 feet long! Along the way, helper organs such as the liver, gallbladder, and pancreas play important roles by producing digestive enzymes and bile to aid in digestion.

Any remaining undigested food then passes into the large intestine, where water and salts are absorbed, forming solid waste known as feces. The large intestine is wider than the small intestine, but if stretched out, it would measure only about 5 feet long.

Finally, any food the body does not need leaves the body as waste through the rectum.

Name: ______________________

Week Six

Day 1

The Digestive System

Read and gather clues from the passage "The Digestive System" to label the Digestive System. One is already labeled for you.

Word Bank

Small Intestine	Stomach	Large Intestine	Esophagus
Mouth	Pancreas	Rectum	Liver

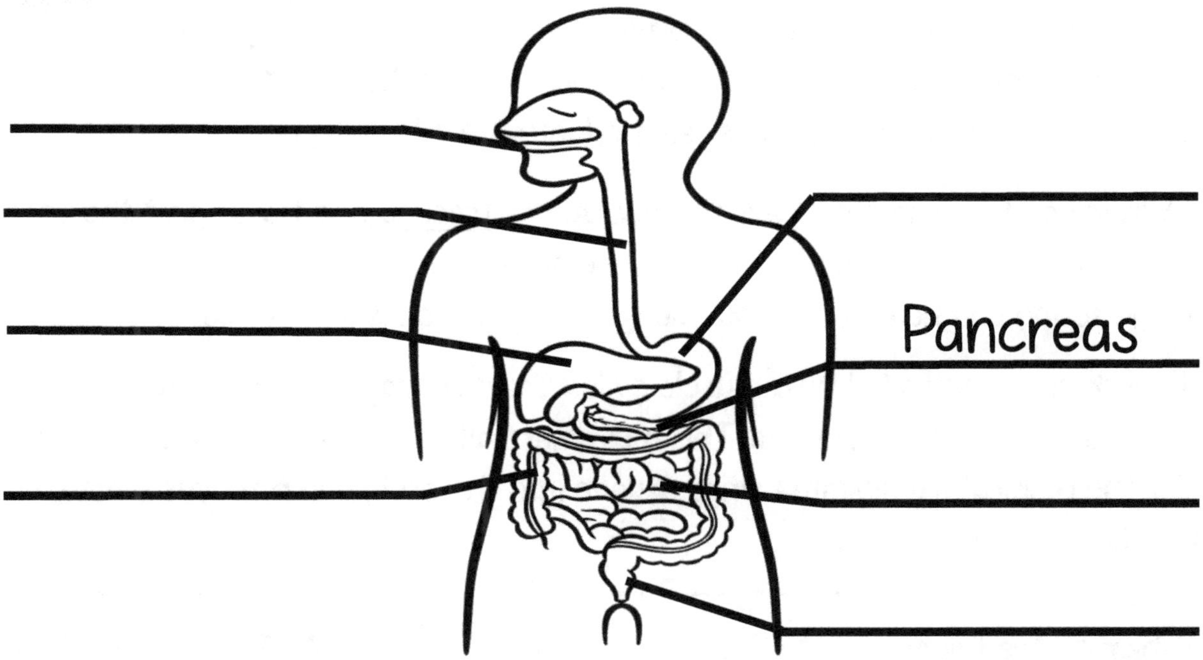

Use the passage to answer the questions.

1. What is the main job of the digestive system?

____

____

2. Where does most of the absorption of nutrients happen in the body?

____

____

Name: ______________________

The Respiratory System

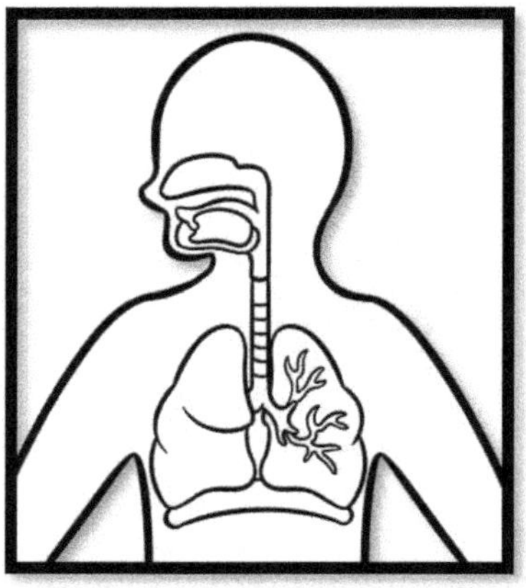

The respiratory system is an important part of our body because it helps us breathe and exchange gases. When we inhale, air goes in through our nose into the nasal cavity or through our mouth, then travels down a tube called the windpipe, also known as the trachea, and into our lungs. Inside our lungs, there are tiny air pockets called alveoli, where the exchange of gases happens. When we breathe in, oxygen enters our lungs and moves into our blood. When we breathe out, carbon dioxide from our body is released into the air.

When you breathe out, the air moves up the trachea and flows past the vocal cords. The vocal cords begin to vibrate as the air moves over them. These vibrations create the sound you hear when you speak or sing, which is your voice.

Breathing is controlled by a muscle called the diaphragm, which moves up and down to help the chest change size. This movement allows air to come in and out of the lungs. So, the respiratory system brings oxygen into the body and removes waste gases so the body can function properly.

---

1. Which of the following is NOT a function of the respiratory system?

a) Carrying oxygen to the body
b) Pumping blood throughout the body
c) Removing waste gases
d) Helping us speak and produce sounds

2. Which of the following is a waste gas that is released when we exhale?

a) Carbon Dioxide
b) Nitrogen
c) Hydrogen
d) Oxygen

3. How does oxygen enter our bodies?

a) Through our skin
b) Through our ears
c) Through our mouth and nose
d) Through our stomach

4. Which muscle helps us breathe by moving up and down?

a) Diaphragm
b) Biceps
c) Quadriceps
d) Hamstrings

Name: ______________________

Week Six

Day 2

Label the Respiratory System

Word Bank: Lungs | Trachea | Nasal Cavity | Diaphragm

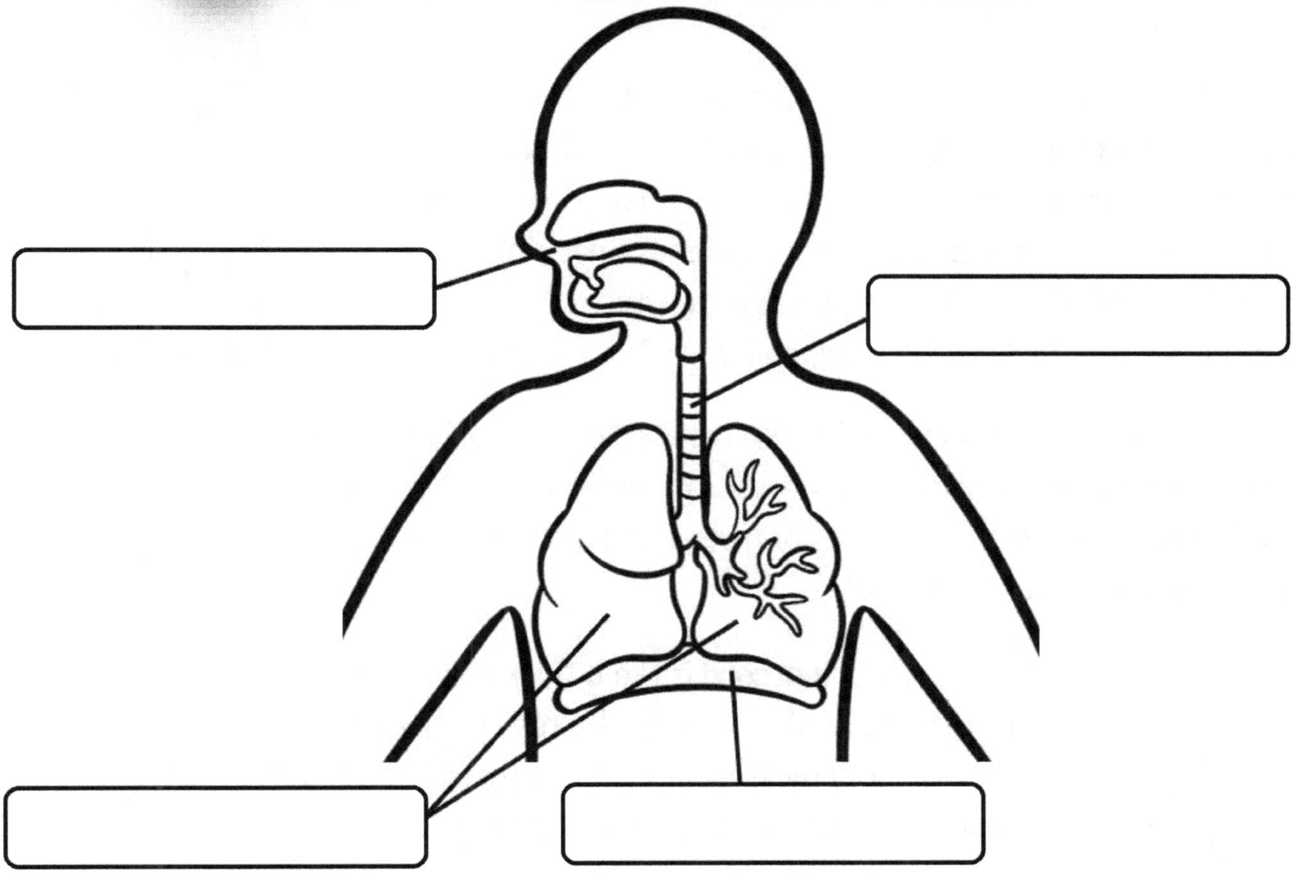

Answer the Questions

Skill Check: Details from the Text	Skill Check: Cause & Effect	Skill Check: Vocabulary
1. What happens in the alveoli? a) Food is broken down b) The exchange of gases takes place c) Blood is pumped through the body d) Sound is made when we speak	**2. Why does the diaphragm move up and down?** a) To help the heart beat faster b) To clean out waste gases c) To help us chew food d) To help the lungs pull air in and push air out	**3. What is the windpipe also called?** a) Trachea b) Bronchi c) Esophagus d) Diaphragm

Week Six

Day 3

Name: ______________________________

The Nervous System

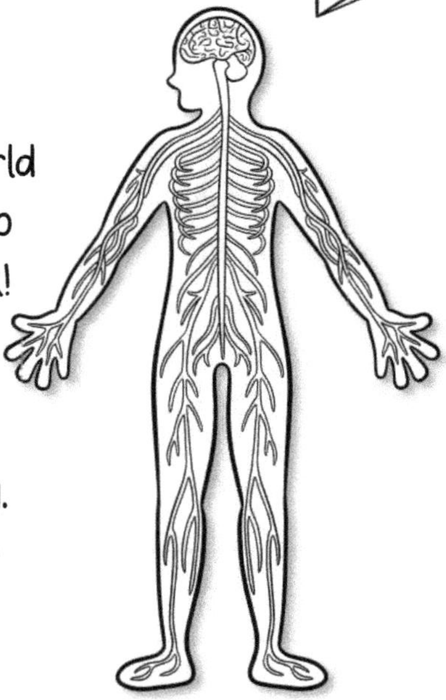

The nervous system is the body's communication network. It helps you think, feel, move, and react to the world around you. Every time you kick a ball, smell cookies, or jump when you hear a loud noise, your nervous system is at work!

The nervous system has two main parts: the **brain** and the **spinal cord**, along with all the **nerves** that branch out from them. The brain is the control center of your body. It helps you think, remember, learn, and make decisions. The largest part of the brain is the **cerebrum**, which is responsible for thinking, learning, and your five senses.

Another important part is the **cerebellum**, which helps you keep your balance and coordinate your movements. A small part of the brain called the **medulla** controls automatic jobs, like breathing and your heartbeat.

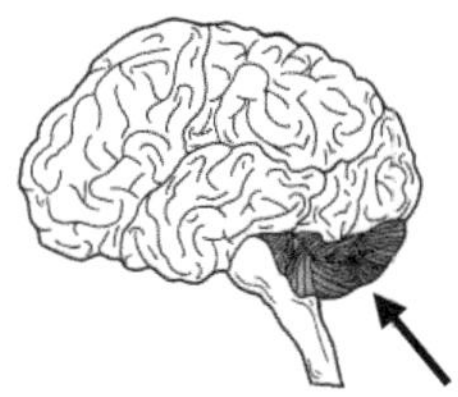

Cerebellum

Spinal Cord

The **spinal cord** is like a thick rope of nerves that connects the brain to the rest of the body. **Nerves** carry messages back and forth between the brain and your muscles, skin, and organs. These messages travel as tiny electrical signals called **impulses**.

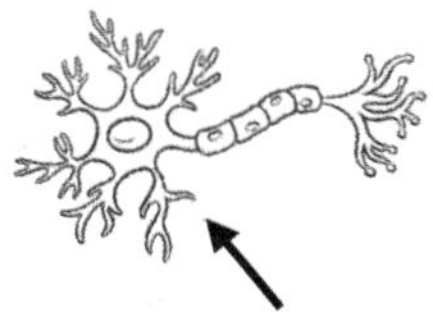

Nerves

Sometimes, your nervous system reacts without you even thinking. This is called a **reflex**. For example, if you touch something hot, your hand pulls away before your brain even has time to think about it! Reflexes protect your body and help keep you safe.

Fun Fact
Some messages in your nervous system travel over 200 miles per hour, faster than many cars can drive!

Name: ______________________________

Week Six

Day 3

Label the Nervous System

Word Bank

Spinal Cord	Cerebrum	Cerebellum	Brain	Nerves

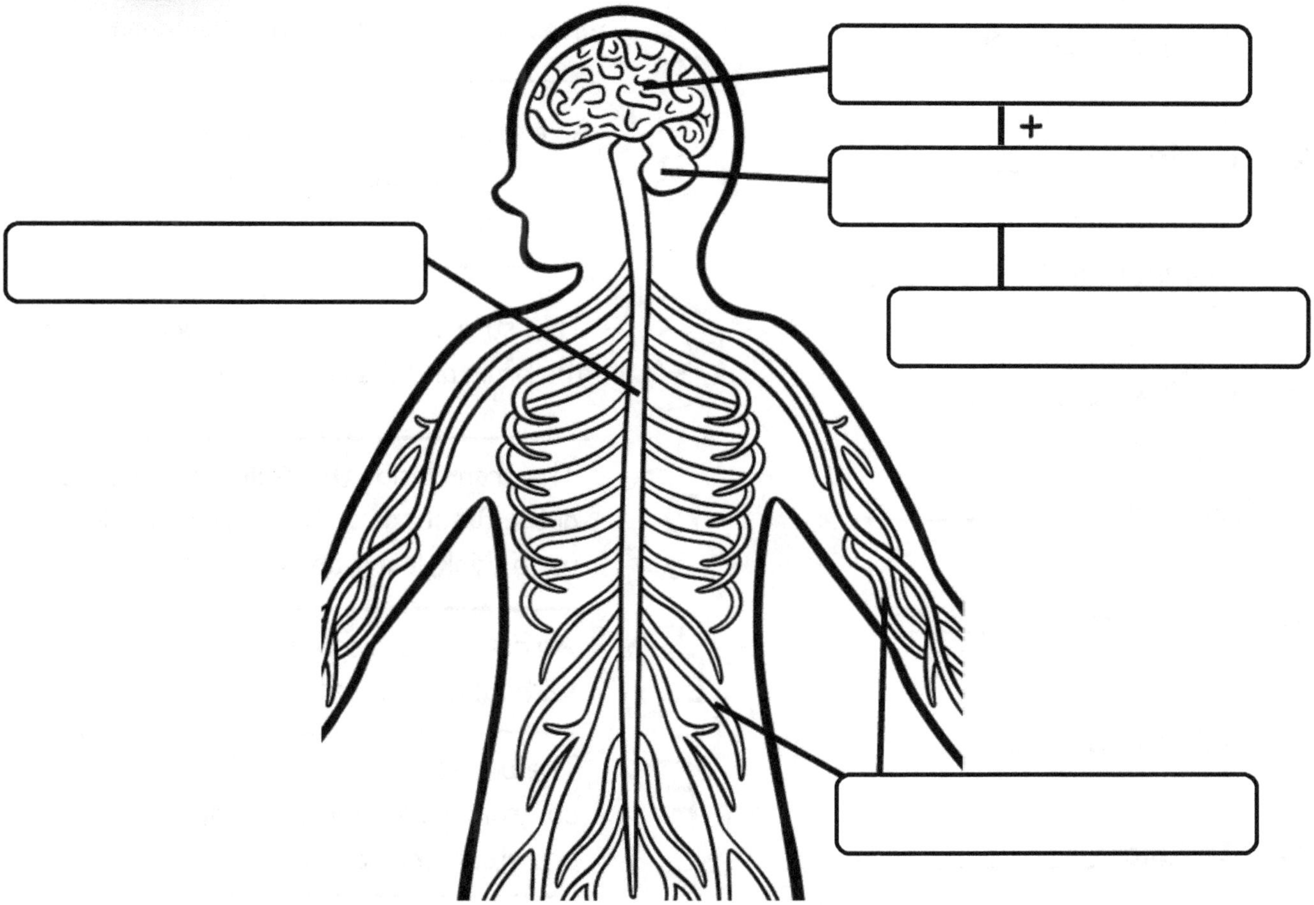

Skill Check: Main Idea / Function	**Skill Check:** Structure & Function	**Skill Check:** Vocabulary in Context
1. What is the main job of the nervous system? a) To break down food b) To help the body communicate and respond c) To move blood around the body d) To clean the lungs	**2. Which part of the brain helps with balance and movement ?** a) Cerebellum b) Cerebrum c) Medulla d) Spinal cord	**3. What is a reflex?** a) A message carried by nerves b) A signal sent from the brain c) An automatic response that protects the body d) A type of muscle in the nervous system

Name: ______________________

The Nervous System

Vocabulary Match

Match each word with its definition. Write the correct letter in the space provided.

1. Nervous System ________
2. Brain ________
3. Spinal Cord ________
4. Reflex ________
5. Cerebellum ________
6. Cerebrum ________
7. Nerves ________
8. Medulla ________
9. Impulses ________

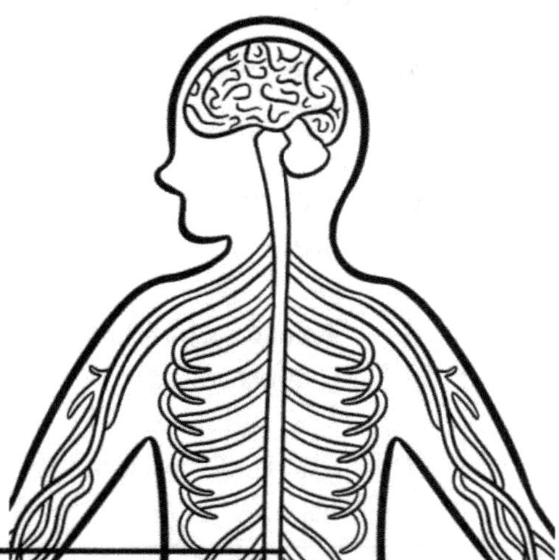

A. The part of your brain in the back that helps with balance and coordinating movement

B. An electrical signal carried by nerve cells when stimulated, also known as "action potential"

C. Thin strands that carry messages between the brain and the rest of the body

D. The organ inside the skull that controls al the functions of the body and helps us to think, learn, and remember

E. An automatic response to a stimulus without conscious thought

F. The biggest part of your brain that controls thinking, learning, and feelings, and stores memories

G. A thick bundle of nerves that carries messages between the brain and the rest of the body

H. The communication network of the body that helps you think, feel, move, and respond

I. Also known as "brain stem". It controls involuntary muscles like breathing, heartbeat, swallowing, and food digestion

Name: ______________________________

Week Six

Day 4

The Skeletal System

Your skeletal system is the strong framework that supports your entire body. It gives you shape, helps you stand tall, and works with your muscles so you can move, run, jump, and even sit up straight. Your bones also protect your important organs. Your skull protects your brain, and your ribcage forms a shield around your heart and lungs to keep them safe. Without your skeleton, your body wouldn't be able to move or stay upright!

Did you know?

Half the bones in your body are in your hands and feet!

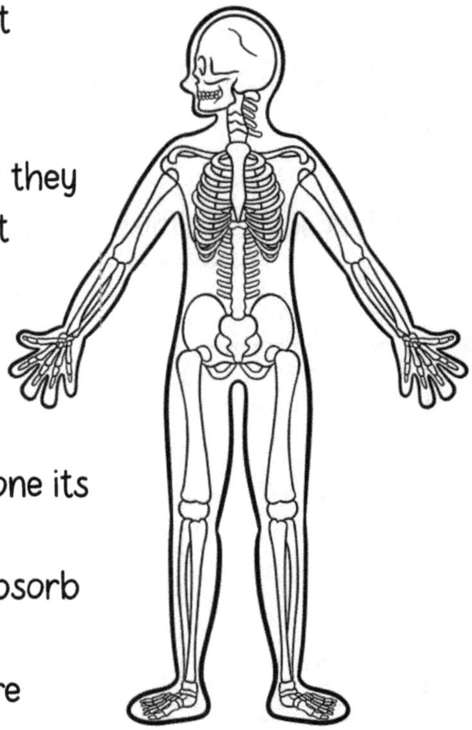

Bones also do something many people don't know: they make blood cells! Red blood cells are made inside the soft material found in the center of many bones.

Bones have three main layers:

- **Compact Bone** – the hard, outer layer that gives a bone its strength.
- **Spongy Bone** – the lighter, flexible layer that helps absorb impact when you move.
- **Bone Marrow** – a jellylike center where blood cells are produced. Some types of marrow also store fat for energy.

Label the Layers of a Bone

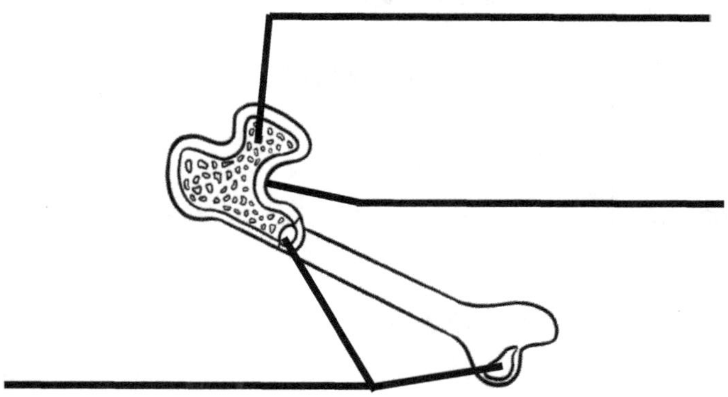

Name: ______________________________

Week Six

Day 4

Joints

Bones can't bend on their own. Joints are the places where bones meet and allow your body to move. There are four main types of movable joints:

•**Ball-and-Socket Joint** – moves in many directions. Example: shoulders and hips.

•**Hinge Joint** – moves back and forth in one direction. Example: elbows and knees.

•**Pivot Joint** – allows twisting and turning. Example: your neck, which helps you turn your head.

•**Gliding Joint** – allows bones to slide past one another. Example: wrists and ankles.

Label the Four Movable Joints

Word Bank

- Ball and Socket
- Gliding
- Hinge
- Pivot

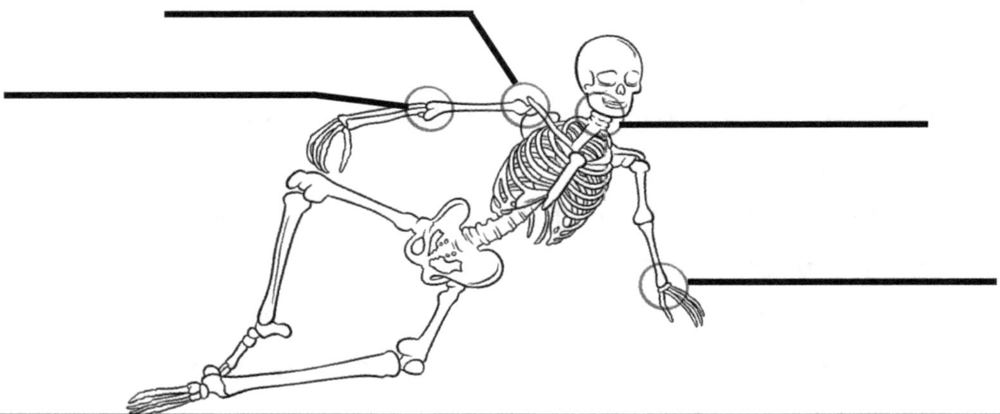

Skill Check: Main Idea	**Skill Check:** Classification	**Skill Check:** Cause & Effect
1. **What is the main job of joints?** a) To store minerals b) To keep bones from touching c) To help the body bend and move d) To protect organs	2. **Which joint allows movement in many directions?** a) Ball-and-socket b) Hinge c) Pivot d) Gliding	3. **Why can't bones move without joints?** a) Bones are too soft b) Joints connect bones and allow them to bend c) Bones would become too flexible to support the body d) Bones would stop protecting the organs

Name: ____________________

The Circulatory System

Did you know?

Your heart beats about 100,000 times every day!

The circulatory system is the body's transportation system. Its main job is to move blood throughout the body. Blood carries oxygen, nutrients, and water to your cells and removes waste materials like carbon dioxide.

The heart is a strong muscle that pumps blood through a network of tubes called blood vessels. There are three main types of blood vessels: arteries, veins, and capillaries. **Arteries** carry blood away from the heart. **Veins** carry blood back to the heart. **Capillaries** are tiny vessels where oxygen and nutrients move into the body's cells.

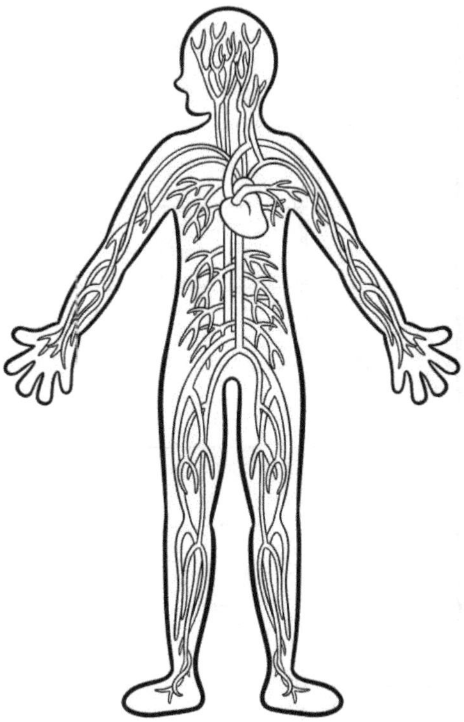

When you exercise, your muscles need more oxygen and energy. Your heart beats faster to pump more blood to your muscles. This helps your body keep moving and working properly.

The circulatory system works closely with the respiratory system. The lungs add oxygen to the blood, and the circulatory system delivers that oxygen to the rest of the body.

Skill Check: Main Idea	**Skill Check:** Cause & Effect	**Skill Check:** Vocabulary
1. **What is the main job of the circulatory system?**	2. **Why does your heart beat faster when you exercise?**	3. **What do arteries do?**
a) To help you breathe	a) To help you breathe faster	a) Carry blood back to the heart
b) To move blood through the body	b) To send more oxygen to your muscles	b) Carry air into the lungs
c) To break down food	c) To cool down the body	c) Carry blood away from the heart
d) To control reflexes	d) To slow your movement	d) Remove waste from the body

Name: ______________________________

Week Six

Day 5

<u>The Circulatory System:</u> Imagine you are running during recess. Explain what happens in your body that helps your muscles keep moving. Use what you learned about the circulatory system in your answer.

Name: ______________________________

Week Six

Extra Practice Page

Name the Body Systems

1

____

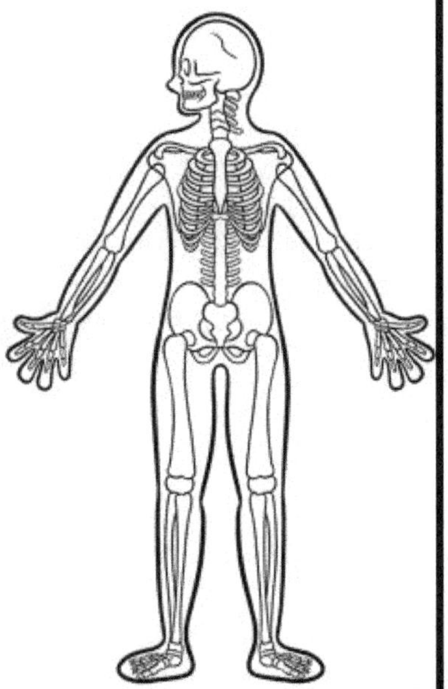

2

____

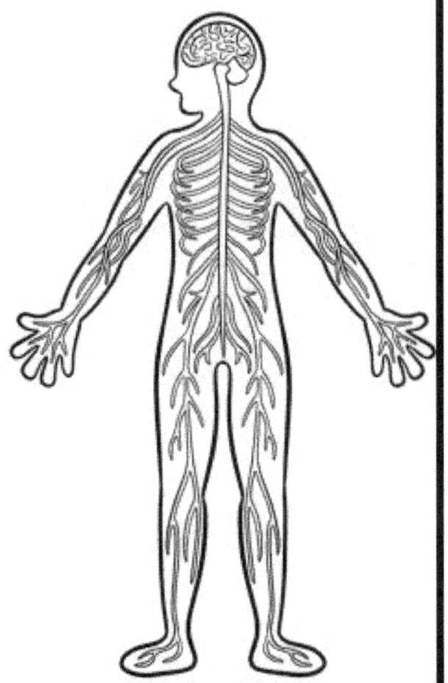

3

____

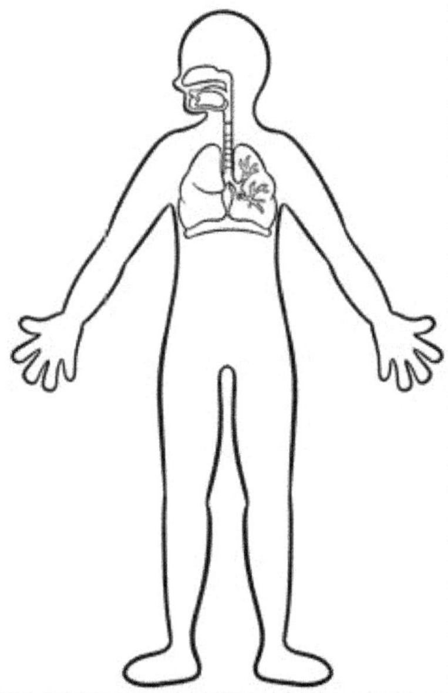

4

____

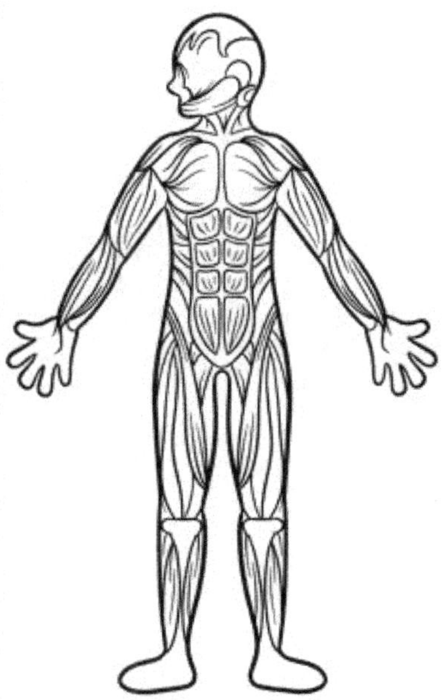

5

____

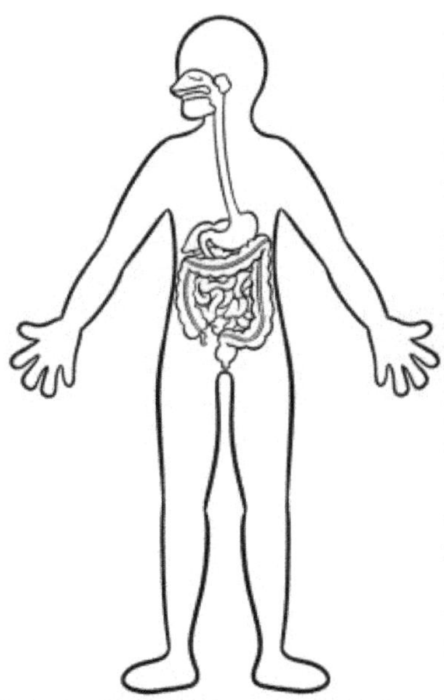

6

____

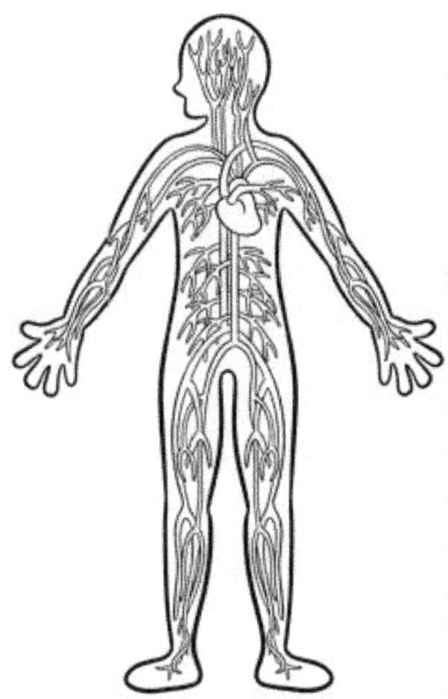

Week Seven

Parts of a Plant

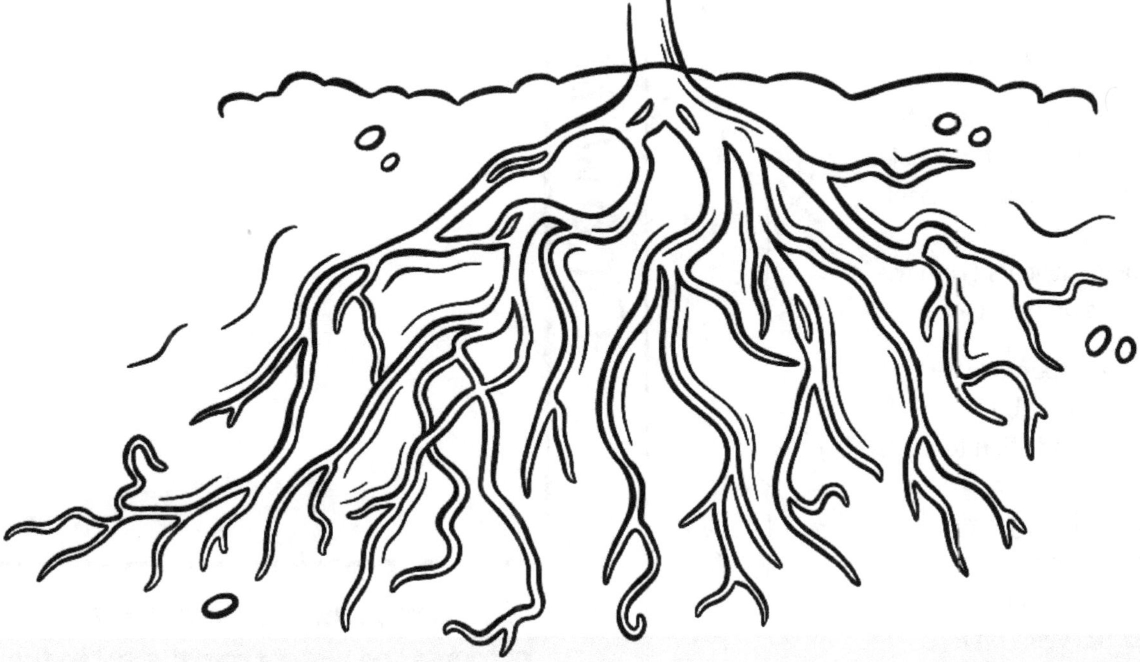

Roots are like the anchor of a plant, keeping it firmly grounded in the soil. They also absorb water and nutrients from the soil, like a straw sucking up a tasty drink!

Week Seven

Day 1

Name: ______________________________

Plant Structures

Plants have different parts called structures that contribute to the plant's life and growth.. Roots, stems, leaves, and flowers are the basic structures you can identify in most plants.

Roots grow in the ground, and they have very important functions. They anchor the plant firmly in place. They also absorb water and nutrients from the soil, which are transported to the rest of the plant. You can find some plants that store the food they make in their roots, like carrots or potatoes.

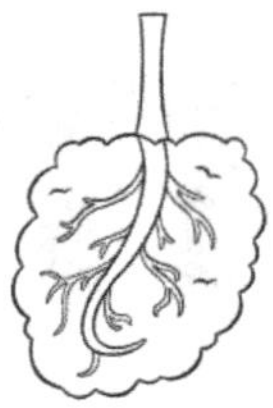

The plant's **stem** connects the roots to the rest of the plant and holds the leaves, flowers, and fruits upright. The stem is also responsible for carrying water and nutrients from the roots to other parts of the plant.

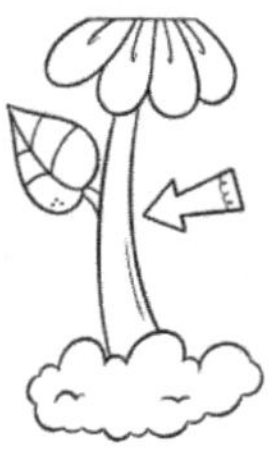

Leaves are responsible for making food for the plant by absorbing sunlight. In addition to sunlight, leaves need water and carbon dioxide to make sugar for the plant. The process of making food for the plant is called photosynthesis. Have you noticed the veins on most leaves? These veins carry the food the leaf makes to the rest of the plant.

Most plants also have brightly colored **flowers**. The beauty and fragrance of flowers attract pollinators such as bees or butterflies. These pollinators help spread pollen from one flower to another, allowing the flowers to produce seeds. After pollination, flowers can grow into **fruits**, which hold and protect the seeds. **Seeds** are very important to plants because new plants can grow from seeds. Without seeds, many plant species would not be able to grow or survive.

1. **Read the passage**
2. **Draw and label a plant using the word bank.**
3. **Answer the questions.**

Word Bank	Leaves	Stem	Fruit	Roots	Flowers

Draw It!

1. **What is responsible for sucking up nutrients and water from the soil to the plant?**

   a) Fruits
   b) Roots
   c) Stem
   d) Flowers

2. **What part of the plant catches sunlight and helps make food?**

   a) Stem
   b) Roots
   c) Buds
   d) Leaf

Week Seven

Day 2

Name: ______________________________

Flower Structures and Their Functions

Read the passage and answer the questions.

Flowers come in many different shapes, sizes, and colors, but most of them have a similar basic structure. **Petals** are often brightly colored and highly scented to attract pollinators such as bees and butterflies.

Sepals are small leaf-like parts found below the petals. Their job is to protect the flower before it blossoms.

The male part of the flower is the **stamen**. The head of the stamen, called the **anther**, is responsible for producing pollen. The **filament** is the stalk that holds the anther and attaches it to the flower.

The female part of the flower is the **pistil**. It is made up of the stigma, style, and ovary. The **stigma** is the head of the pistil that receives the pollen which will begin the process of fertilization.

The **style** is the stalk of the pistil where the stigma sits. The **ovary** is the base of the pistil and is usually found at the bottom of the flower. This holds the **ovules** (eggs), which become seeds after fertilization.

Did you know?

Some flowers can pollinate themselves, but many need insects or wind to help!

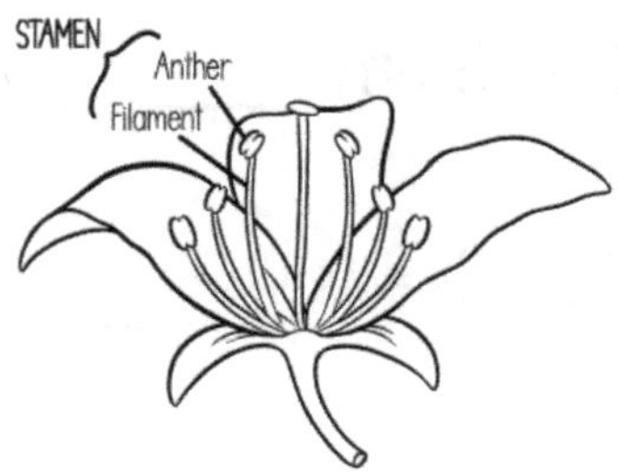

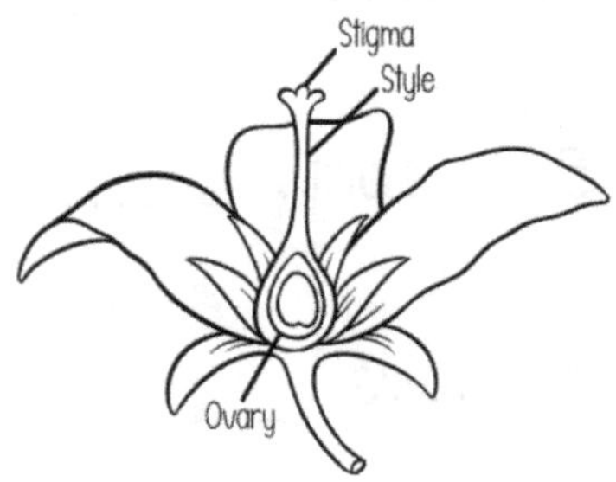

1. Why are petals often brightly colored and highly scented?

a) To protect the flower from predators
b) To store food for the plant
c) To provide structural support
d) To attract pollinators

2. What is the male part of the flower called?

a) Stamen
b) Pistil
c) Sepal
d) Patel

3. What is responsible for producing pollen?

a) Stigma
b) Anther
c) Ovary
d) Style

4. What is the female part of the flower called?

a) Stamen
b) Patel
c) Pistil
d) Sepal

Name: ______________________________

Parts of a Plant

Vocabulary Match

Match each word with its definition. Write the correct letter in the space provided.

1. Leaves ________
2. Stem ________
3. Seeds ________
4. Roots ________
5. Sepals ________
6. Buds ________
7. Stamen ________
8. Petals ________
9. Pistil ________

A. The part of a plant that can grow into a new plant

B. The male part of a flower that produces pollen

C. Protects the flower before it opens

D. Grows on a stem and absorbs sunlight

E. The female part of a flower that creates seeds

F. Colorful parts of a flower that attracts pollinators

G. Grows underground, holds the plant steady, and absorbs water and nutrients from the soil

H. Little bumps on the stem that grow into leaves, flowers, and branches

I. Holds up the plant and carries water and nutrients from the roots to the leaves, flowers, and fruit

Week Seven

Day 4

Name: ______________________________

Label the parts of a flower using the word bank.

WORD BANK

Ovary	Stem	Anther	Style
Sepal	Filament	Ovule	Stigma

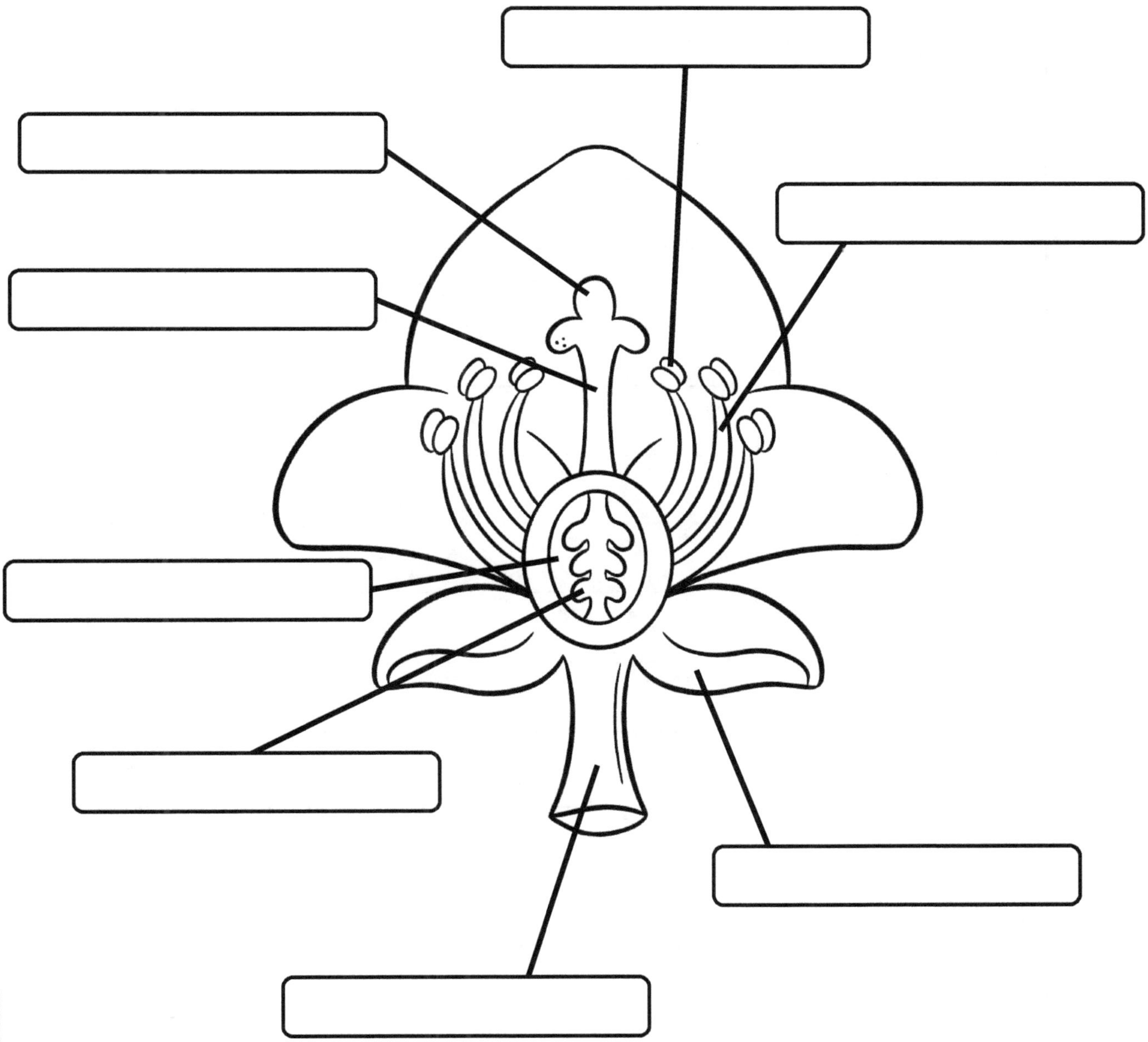

Challenge:
Circle the part that makes pollen.
Draw a star on the part that receives pollen

Name: ______________________

Some plants have thorns or spines on their stems or leaves. Why do you think these plants developed thorns or spines? How do these features help plants survive?

Week Eight

Plant Life Cycle

Did you know that bamboo is the fastest growing plant in the world? In fact, bamboo can grow up to 35 inches in just one day!

Name: ______________________

The Journey of a Plant

It all starts with a tiny seed. When a seed is planted in the soil and absorbs water while being warmed by sunlight, it begins to germinate. **Germination** is when the seed starts to sprout and grow roots.

As the seed grows, it develops into a **seedling**. The seedling pushes its way up through the soil, and its tiny leaves begin to unfold. The roots grow deeper into the soil to anchor the plant and absorb water and nutrients. With time, the seedling grows into a **mature plant**. It continues to grow taller and wider, producing more leaves, stems, and branches. The plant's main job is to make its own food through a process called **photosynthesis**, using sunlight, water, and carbon dioxide.

When a plant reaches maturity, it may produce flowers. **Flowers** are not only beautiful, but also essential for the plant's reproduction. After **pollination**, which happens when pollinators like bees transfer pollen between plants, the flower begins to produce seeds. The **seeds** contain the genetic information needed to grow into new plants. Some plants spread their seeds by wind, water, or animals. This is called **seed dispersal**.

The cycle then begins again, with seeds germinating and growing into new plants, continuing the circle of life for plants.

Number the steps of the plant life cycle from 1-6, with 1 being the first stage and 6 being the last.

Flowers begin to bloom on the plant.

____

The seedling grows above the soil.

____

The seed absorbs water ______________

Roots start growing out of the seed.

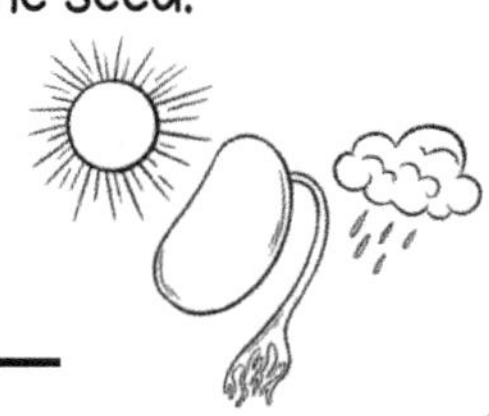

____

The flowers are pollinated to make seeds.

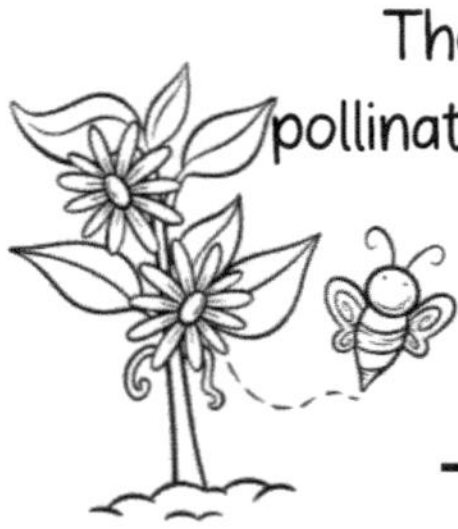

____

The plant is growing more stems, leaves, and branches.

____

Week Eight

Day 2

Name: ______________________________

Describe each term in your own words and illustrate it with a drawing.

	Explain	Picture
Seed		
Germination		
Seedling		
Young Plant		
Mature Plant		
Pollination		

Name: ____________________

Plant Life Cycle

Vocabulary Match

Match each word with its definition. Write the correct letter in the space provided.

1. Life Cycle ______
2. Seed ______
3. Germination ______
4. Seedling ______
5. Embryo ______
6. Pollination ______
7. Pollen ______
8. Seed Dispersal ______
9. Reproduce ______

A — The process that happens when pollen is transferred to help a plant make seeds

B — A small part of a plant from which a new plant can grow

C — A powder made by flowers that helps plants make seeds

D — When a living thing makes new living things of the same kind

E — When a seed begins to sprout and grow roots

F — The stages a plant goes through during its life

G — A tiny baby plant inside a seed

H — The way seeds are spread away from the parent plant

I — A young plant that has just grown above the soil

Week Eight

Day 4

Name: ______________________________

Under each picture, write what you see. Use the word bank.

WORD BANK

Young Plant, Germination, Seed, Pollination, Seedling, Mature Plant

____

____

____

____

____

____

Answer the multiple-choice questions.

1. What do seeds need to begin germination?

a) Only sunlight
b) Air, water, and soil
c) Warm temperatures and nutrients
d) Pollinators and wind

2. What do the flowers produce through pollination?

a) Fruits
b) Leaves
c) Seeds
d) Roots

3. When the first sign of plant life appears above the soil, it is called a _____.

a) Flower
b) Fruit
c) Seedling
d) Root

4. What is the stage in the plant life cycle where a new plant begins to grow from a seed?

a) Pollination
b) Seedling
c) Mature Plant
d) Germination

PARTS OF A PLANT

Week Eight

Day 5

Name: ______________________________

Why is pollination an important step in the plant life cycle? How does pollination help plants make new seeds?

Week Nine

Photosynthesis

Did you know that one large tree can provide a day's supply of oxygen for up to four people?

Week Nine

Day 1

Name: ______________________________

Understanding Photosynthesis

It would be weird to see a plant eating food the same way we do, don't you think? The way a plant gets the energy it needs to grow and stay alive is completely different from humans and animals. We call the process plants use to make their own food **photosynthesis**.

Photosynthesis is the process plants use to take the energy from the sun and use it to turn carbon dioxide and water into food. The food they make is called **glucose,** which is a type of sugar.

The three basic things plants need to stay alive are sunlight, water, and carbon dioxide. Without any of these three things, plants will die. If all plants died, we would die too because plants release oxygen, and we need oxygen to breathe. Plants are a major source of oxygen on Earth. While we breathe in oxygen, we also breathe out carbon dioxide. So, in a way, we are helping plants too!

Plants get energy from sunlight using a green chemical in their leaves called **chlorophyll**. Chlorophyll absorbs the sun's energy, and it is also what gives leaves their green color.

Plants get water and minerals from the soil through veins called **xylem,** which carry them to the leaves. The next time you water a plant, make sure you water the soil! Gardeners use fertilizer to give plants extra nutrients because the soil does not always have enough nutrients, such as nitrogen, phosphorus, and potassium. Some fertilizers are made from chemicals, while others are made from natural materials, like cow poop. That's right – poop is full of nutrients for plants!

1. **True or False:**
   Plants eat food the same way humans do.

   ____

2. What is the process plants use to make their own food called?

   ____

3. How do plants make their own food?

   ____

   ____

   ____

   ____

4. What is the food that plants make called?

   ____

5. What are the three basic things plants need to stay alive?

   ____

   ____

6. Why would we die if all the plants in the world died?

   ____

   ____

   ____

7. What absorbs the sun's energy?

   ____

8. How do plants get water and minerals?

   ____

   ____

Name: ______________________

Steps of Photosynthesis

Imagine you are exploring a vibrant garden filled with colorful flowers and tall trees. Have you ever wondered how these plants grow so tall and lush? It is all thanks to the fascinating process called **photosynthesis**.

It is a sunny day in the garden, and sunlight is what plants need for photosynthesis to work. But how do they capture it? That's where chlorophyll comes in. **Chlorophyll** is a green pigment found in the leaves of plants that absorbs energy from the sun.

Leaves have tiny pores called **stomata**. Through these tiny openings, plants take in something very important: carbon dioxide. Just like we need oxygen to breathe, plants need carbon dioxide to make food. Meanwhile, down below the ground, the plant's roots are busy getting water from the soil. The water travels up through the stem, all the way to the leaves.

Now the leaves have sunlight, carbon dioxide, and water. Inside the leaves, water and carbon dioxide combine with the energy from sunlight to make a special kind of sugar called **glucose**. The plant uses some of this glucose for energy to grow and stay healthy. During this process, the plant also releases oxygen into the air, which we need to breathe.

Plants are not wasteful. They store any extra water and glucose they don't need right away in special compartments called **vacuoles**. This way, they have reserves for later, just in case.

Number the steps of photosynthesis from 1-6, with 1 being the first stage and 6 being the last.

Carbon dioxide enters the leaves through stomata, and water enters through the roots. ______	Sunlight shines on the leaves and is absorbed by chlorophyll. ______	The plant uses the glucose and releases oxygen into the air. ______
Water and carbon dioxide combine with energy to make glucose and oxygen. ______	Extra water and glucose are stored in vacuoles for later use. ______	Water travels up through the stem to reach the leaves. ______

Week Nine

Day 3

Name: ______________________________

Photosynthesis

Vocabulary Match

Match each word with its definition. Write the correct letter in the space provided.

1. Photosynthesis ________
2. Carbon dioxide ________
3. Oxygen ________
4. Glucose ________
5. Chlorophyll ________
6. Xylem ________
7. Stomata ________

(A) A type of sugar plants make for food

(B) Tiny holes in leaves that let gases move in and out of the plant

(C) A gas that plants absorb from the air to make food

(D) A green substance in the leaves of plants that absorbs sunlight

(E) The process plants use to take energy from sunlight and turn water and carbon dioxide into food

(F) The gas plants release into the air that people and animals need to breathe

(G) Plant veins that carry water from the roots to all parts of a plant

Week Nine

Day 4

Name: ______________________________

Photosynthesis Process Labeling

Use the word bank and the arrows to label the diagram showing the photosynthesis process.

WORD BANK Carbon Dioxide Water Oxygen Energy

PHOTOSYNTHESIS

Week Nine

Day 5

Name: ______________________________

What would happen if plants could not make their own food through photosynthesis? Explain how this would affect oxygen in the air and living things that depend on plants.

Exploring Earth Science

Week One

Rocks and Minerals

Some rocks on Earth come from space! These are called meteorites and they provide clues about our solar system.

Week One

Read the passage to complete the tasks for the next few days.

Types of Rocks

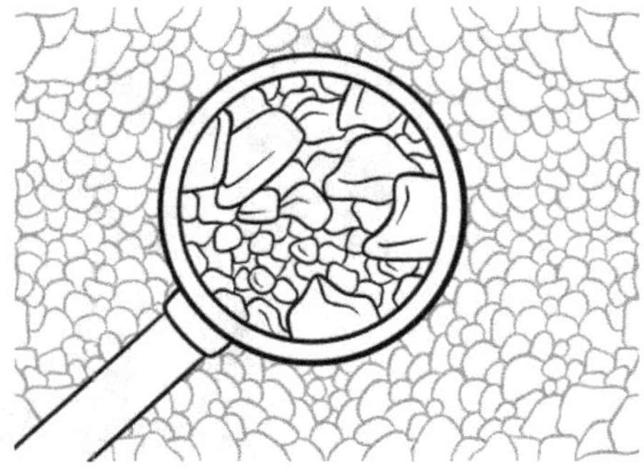

Rocks are everywhere. When you go outside, look around and notice how many different kinds of rocks you can find. Every rock is unique. Rocks are made of minerals, which are natural materials found in the Earth. Each mineral looks and feels different and has its own color and hardness. Examples of minerals include quartz, salt, and gold. These minerals join together to form the many different rocks we see.

There are three main types of rocks: igneous, sedimentary, and metamorphic.

Igneous rocks form when molten rock from inside the Earth cools and hardens. When a volcano erupts, it releases extremely hot molten rock called magma. When magma reaches the Earth's surface and cools, it becomes igneous rock. Granite is an igneous rock with large crystals that forms when magma cools slowly beneath the Earth's surface. When magma cools quickly after a volcanic eruption, it forms igneous rocks like basalt, which have very small crystals.

Examples of Igneous Rocks

GRANITE

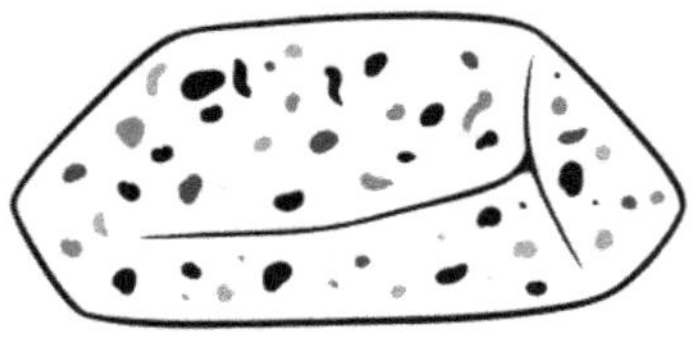

Has large crystals

BASALT

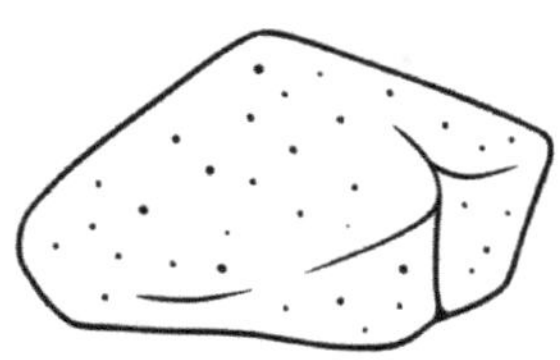

Has small crystals

OBSIDIAN

Smooth, glassy texture

Sedimentary rocks form over many years as layers of sediment are pressed and cemented together. Sediments are small pieces of rocks, minerals, and organic materials. These sediments can settle at the bottom of lakes, oceans, and rivers or collect on land. Common sedimentary rocks include sandstone, which is made from sand particles, and limestone, which forms from shells and remains of marine organisms. Sedimentary rocks often contain fossils that give us clues about Earth's past.

Week One

Examples of Sedimentary Rocks

SANDSTONE

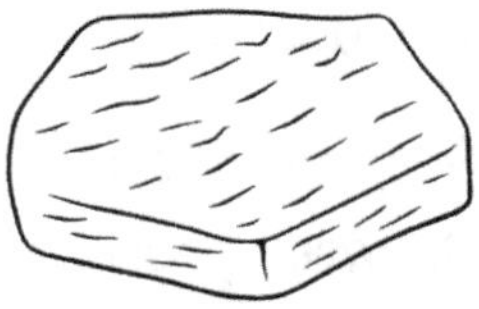

Grainy and layered
Often found in deserts and beaches

LIMESTONE

Smooth and contains visible fossils

SHALE

Has a fine texture often found in lakes and deep ocean floors

Metamorphic rocks form when rocks are changed by great heat and pressure. These conditions are found deep within the Earth's crust. Metamorphic rocks are usually formed from other types of rocks. For example, marble is a metamorphic rock that forms when limestone is exposed to high heat and pressure.

Examples of Metamorphic Rocks

SLATE

Very fine texture

MARBLE

Visible tiny sparkling crystals

GNEISS

Has bands or stripes of different colors

The next time you find a rock, think about how it formed and how it ended up where you found it. Every rock has a story to tell.

Name: ______________________________

Week One

Day 1

Types of Rocks

After reading the passage "Types of Rocks", answer the following questions.

1. **Which type of rock is formed from the cooling and hardening of magma?**
   a) Sedimentary
   b) Igneous
   c) Metamorphic

2. **Sedimentary rocks are typically formed by which process?**
   a) Melting and solidification
   b) Compaction and cementation
   c) Heat and pressure
   d) None of the above

3. **Which type of rock often contains visible fossils?**
   a) Sedimentary
   b) Igneous
   c) Metamorphic

4. **Which type of rock is formed by great heat and pressure?**
   a) Sedimentary
   b) Igneous
   c) Metamorphic

5. **What are sediments?**

6. **Where are metamorphic rocks usually found?**

7. **How is marble formed?**

8. **What are rocks made up of?**

Week One

Day 2

Name: ______________________

The Rock Cycle

After reading the passage, label the rock cycle.

The rock cycle explains how rocks change from one type to another over time. The cycle begins with **igneous rocks**, which form when hot molten rock, called magma, cools and hardens. Over time, these rocks can break down into small pieces called sediments through weathering and erosion.

The sediments are then carried by wind, water, or ice and eventually settle in layers. These layers often form at the bottom of rivers, lakes, or oceans. Over time, the layers become buried, compacted, and cemented together, forming **sedimentary rocks**.

Sedimentary rocks can be pushed deep into the Earth by movements in the Earth's crust. Deep underground, they are exposed to intense heat and pressure, which changes them into **metamorphic rocks**. If these rocks become hot enough to melt, they turn back into magma. When the magma cools again, new igneous rocks form, and the rock cycle begins again.

LABEL THE ROCK CYCLE

Word Bank	Sediments	Magma	Metamorphic	Sedimentary	Igneous

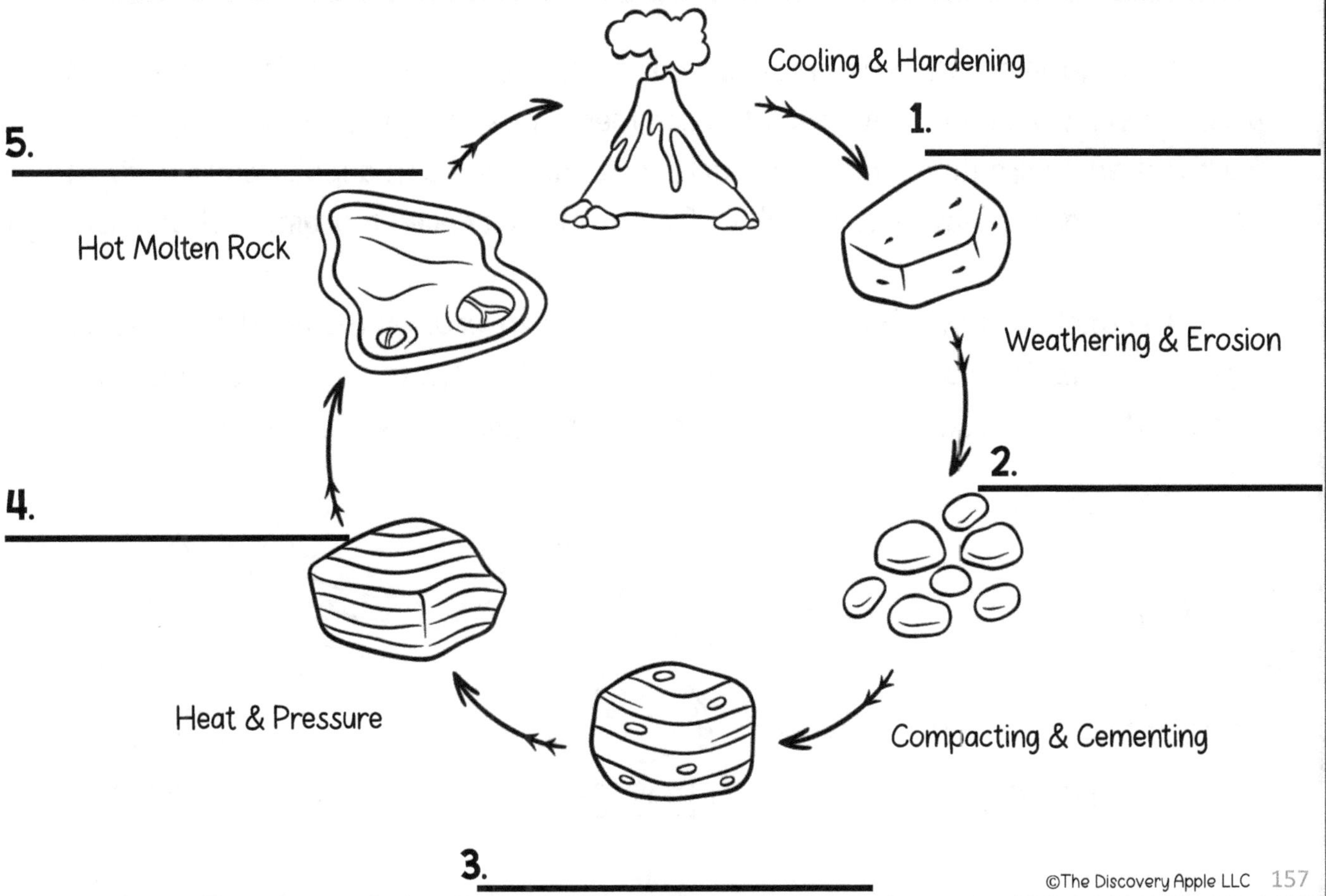

Name: ____________________

Week One

Day 3

Weathering, Erosion, and Deposition

Weathering, erosion, and deposition are processes that shape the Earth's surface. **Weathering** is the process by which rocks break down into smaller pieces. There are two main types of weathering: physical and chemical. Physical weathering is also called mechanical weathering. Mechanical weathering breaks rocks apart without changing what they are made of. For example, when large waves crash against a rocky shoreline, the rocks slowly break into smaller pieces. **Chemical weathering** happens when rocks and minerals are dissolved or broken down through chemical reactions. For example, when water enters cracks in a rock, it can dissolve the minerals inside.

Erosion is the process that moves the small pieces of rock, called sediments, from one place to another. Erosion can happen through wind, water, ice, or gravity. For example, rivers can carry sediments downstream, wind can blow sand across deserts, and glaciers can push rocks along as they move.

Deposition occurs when moving sediments settle in a new place. For instance, when a river slows down, it drops the sediments it is carrying, forming layers along the riverbed or creating landforms such as deltas. Wind can deposit sand to form dunes, and glaciers leave behind piles of rock and soil called moraines when they melt.

In conclusion, weathering breaks rocks down, erosion moves the pieces, and deposition lays them down in new locations. Together, these processes constantly reshape the Earth's surface and help form sedimentary rocks over time.

WEATHERING

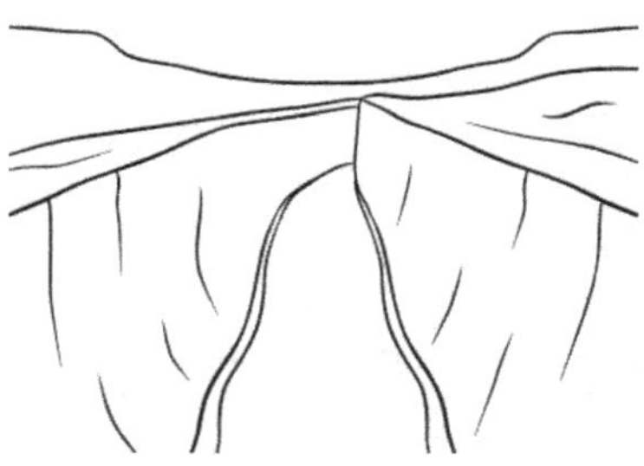

EROSION

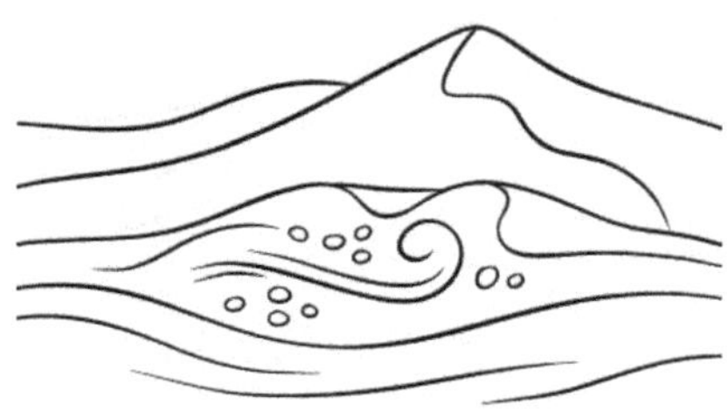

DEPOSITION

Name: ______________________

Week One

Day 3

Weathering, Erosion, and Deposition

Does each statement refer to weathering, erosion, or deposition? Write your answer below each statement.

1. It includes the movement of particles and surface materials.

   ____

2. Rocks are dropped off and settle in a new place.

   ____

3. You can classify it as chemical or mechanical.

   ____

4. It is the process where rock is broken down into small pieces.

   ____

5. The formation of a delta at the mouth of a river

   ____

6. Wind can pick up and carry loose particles away.

   ____

7. Sediments are added to a landform.

   ____

8. It can dissolve minerals in rocks.

   ____

9. It is the wearing away of land.

   ____

Name: ______________________________

Week One

Day 4

Vocab Match

Match each word with its definition. Write the correct letter in the space provided.

1. Rock ________
2. Minerals ________
3. Igneous Rocks ________
4. Sedimentary Rocks ________
5. Metamorphic Rocks ________
6. Magma ________
7. Sediment ________
8. Weathering ________
9. Erosion ________
10. Rock Cycle ________

A. Hot molten rock found deep below the Earth's surface

B. Rocks formed when layers of sediment are compacted and cemented together over time

C. The movement of rock pieces and sediments from one place to another

D. A hard, solid material made up of one or more minerals that makes up the Earth's surface

E. The process that breaks down rocks into smaller pieces through physical or chemical forces

F. Rocks that have been changed from other rocks by heat and pressure

G. Rocks formed when magma cools and hardens

H. Small pieces of broken-down rock and soil

I. A continuous process in which rocks are formed, changed, broken down, and re-formed over time

J. Natural substances found in the Earth, such as salt, quartz, and gold

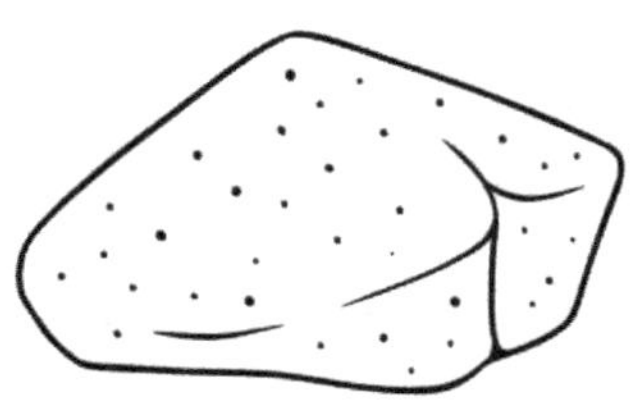
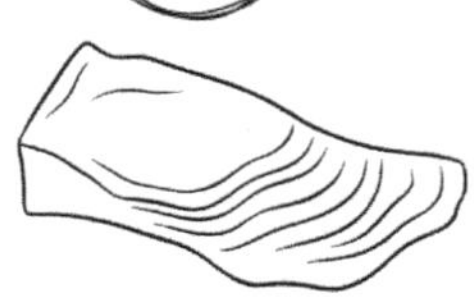

Week One

Day 5

Imagine you are a geologist studying a new rock formation. How could you use your knowledge of the rock cycle to explain the history of these rocks? What evidence would you look for to understand how the rocks formed and changed over time?

Week Two

Soil

A single shovel of soil can contain more living organisms than there are people on Earth! This includes bacteria, fungi, insects, and worms.

SOIL

Name: ____________________

Week Two

Day 1

Understanding Soil

Soil forms on the Earth's surface where plants grow. It is made up of small pieces of rock, sand, clay, dead plants, animal remains, and fungi. Soil forms slowly and can take up to 1,000 years for just one inch of soil to develop. Soil is created through the weathering of rocks, which are broken down by ice, frost, wind, and water.

Soil plays an important role in life on Earth. It helps clean water by acting as a natural filter and releases gases, such as carbon dioxide, into the air. Plants also need soil to grow because it provides nutrients. A scientist who studies soil is called a **pedologist**. Pedologists examine different types of soil and work to keep soil healthy.

here are four main types of soil: sand, silt, clay, and loam. One of the most important properties of soil is its texture, which describes whether the soil is mostly sand, silt, clay, or a mixture of these materials.

Examine the properties of each soil type listed below.

SAND

- The largest particles found in soil
- Feels rough and gritty when rubbed
- Does not contain many nutrients
- Allows water to drain easily

SILT

- Smooth to the touch
- Leaves dirt on your skin when rubbed between fingers
- Retains water longer than sand
- Fairly fertile
- Drains poorly

CLAY

- The smallest particles found in soil
- Holds water well
- Sticky when wet and smooth when dry
- Rich in nutrients for plant growth
- Can become hard and compact in dry conditions

LOAM

- A mixture of sand, silt, and clay
- The best soil for growing plants
- Dark-colored, soft, and crumbly
- Holds water well but still drains properly

Name: ____________________________

Week Two

Day 1

After reading the passage "Understanding Soil" and examining the 'properties of soil chart', answer the following questions.

1. How does soil play an important role to life on Earth?

a) Soil helps clean our water acting as a filter
b) Plants get nutrients from soil to grow
c) Soil releases gases such as carbon dioxide into the air
d) All of the above

2. Which soil type is the best for growing plants?

a) Clay
b) Silt
c) Sand
d) Loam

3. What is soil primarily formed from?

a) Animal remains
b) Weathering of rocks
c) Plant roots
d) Fungi

4. What is a scientists who specifically studies soil called?

a) Geologist
b) Botanist
c) Pedologist
d) Ecologist

Let's Reflect!

5. Why do you think it takes so long for soil to form?

____

____

____

____

6. If you had a garden, which type of soil would you prefer and why?

____

____

____

____

Name: ____________________

Week Two

Day 2

Identify the Soil Type

Read each statement below and write the correct soil type (sand, silt, clay, or loam) each statement is referring to.

1. Feels smooth and powdery when dry and slippery when wet

____

2. The best type of soil for gardening and farming

____

3. Feels gritty and course to the touch

____

4. Its particles are the smallest and finest among the different types of soil.

____

5. Feels sticky and molds easily when wet, but it becomes hard and compact when dry

____

6. Has the largest particles

____

7. Contains a good mix of the three soil materials

____

8. Retains water, drains poorly, and is fairly fertile

____

9. It could turn hard and compact in the summer

____

Name: ______________________________

Soil Layers

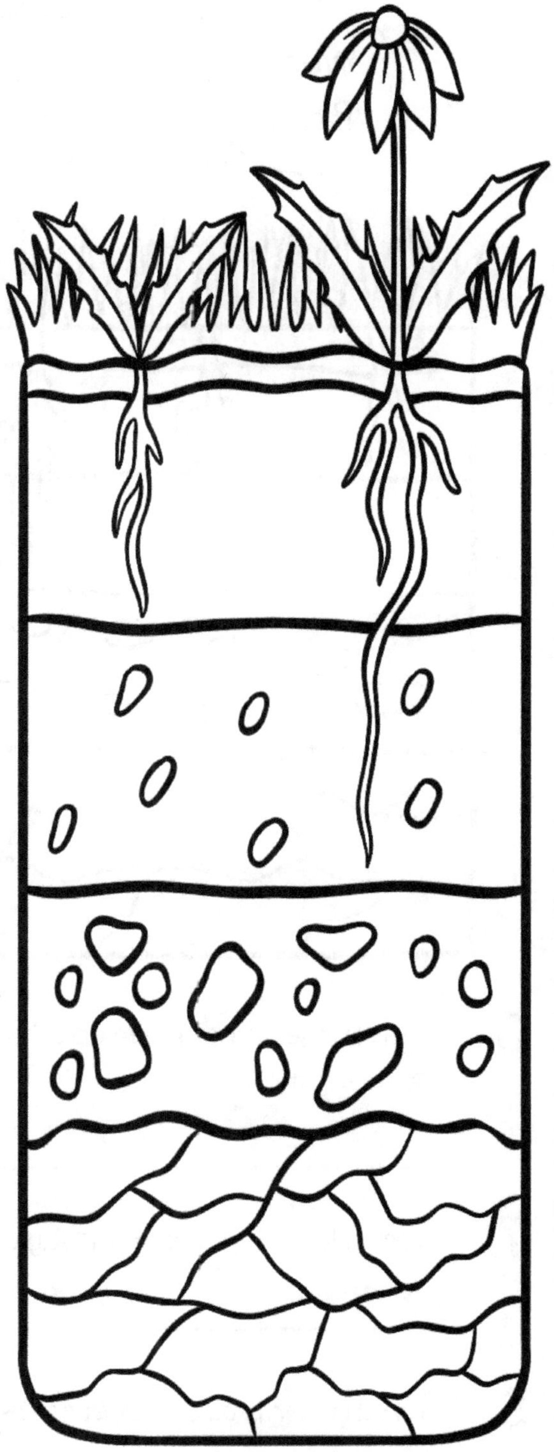

Soil is made up of many layers called **horizons**. Each layer has its own characteristics that help determine how the soil is used and why it is important. When all the layers are viewed together, they form what is called a **soil profile**.

Humus is the top layer of soil. It is dark in color and rich in nutrients. Humus is made up of living and decomposed materials such as leaves, plants, and insects. This layer is also known as the organic layer.

Just below the organic layer is the **topsoil layer**. This layer is also dark and fertile. It is a mixture of organic material and minerals and is the best layer for growing plants. Most plant roots grow in topsoil. This layer is usually fairly thin, about 5–10 inches thick.

Beneath the topsoil is the **subsoil layer**, which is lighter in color and more compact. This layer collects minerals, such as iron and clay, that move down from the layers above. Subsoil is less fertile than topsoil, but it is important for storing water and nutrients.

Below the subsoil is the **parent material**. This layer is made mostly of large rock fragments and pieces of weathered rock. Very few plant roots grow in this layer. It is called parent material because the upper soil layers develop from it over time.

The deepest layer is called **bedrock**, and it is not soil. Bedrock is made of solid rock, such as granite, basalt, limestone, sandstone, or quartzite.

Name: ____________________

Soil Layers

Label the soil layers and answer the questions below.

Word Bank: Subsoil Bedrock Humus Parent Material Topsoil

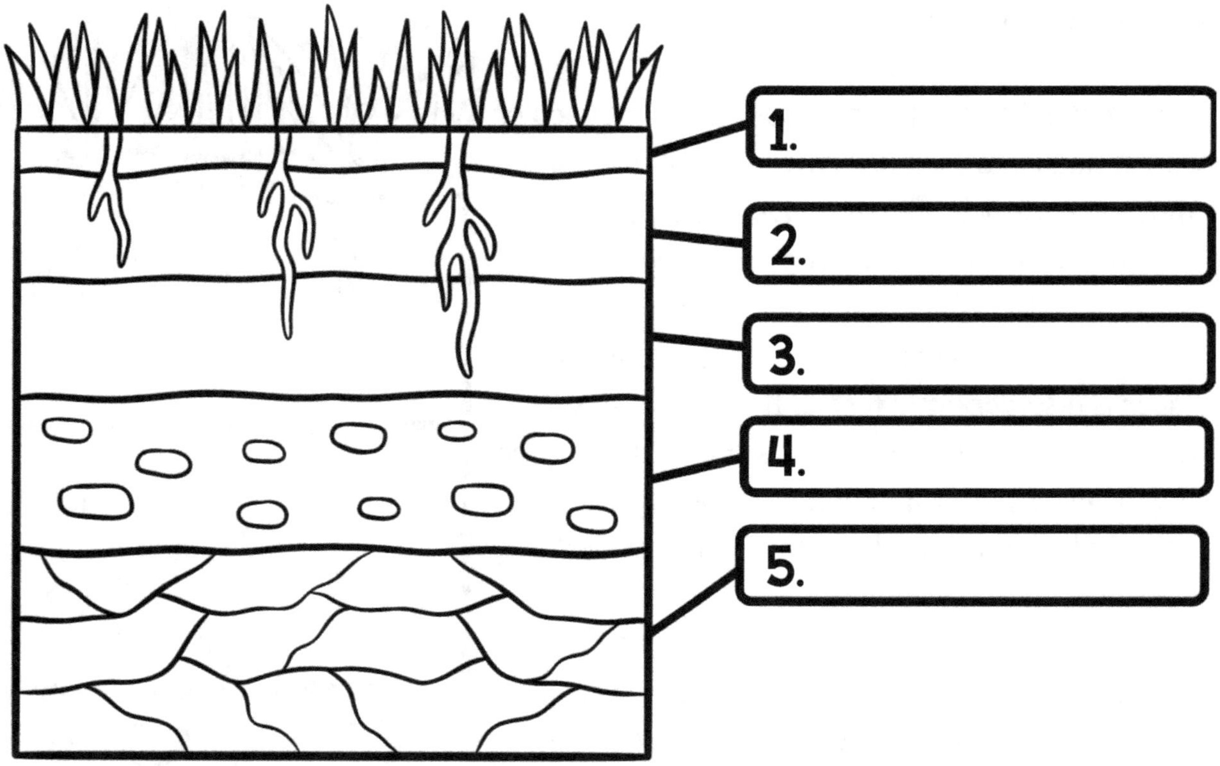

6. Which layer is not made up of soil? (It consists of solid rock, such as granite, basalt, or limestone.)

____

7. Which layer is made up mostly of clay and iron?

____

8. Which layer is also known as the organic layer?

____

Name: ______________________________

Week Two

Day 4

Vocab Match

Match each word with its definition. Write the correct letter in the space provided.

1. Loam ________
2. Bedrock ________
3. Sand ________
4. Clay ________
5. Soil ________
6. Humus ________
7. Silt ________
8. Topsoil ________
9. Subsoil ________
10. Nutrients ________

A. The solid layer of rock found below the soil

B. Material found on the Earth's surface made up of rock particles, sand, clay, dead plants, and animal remains

C. The top layer of soil made up of broken-down plant and animal material

D. Soil with the smallest particles

E. Soil made up of small, smooth particles that are smaller than sand but larger than clay

F. Soil with the largest particles

G. Soil that is a mixture of sand, silt, and clay

H. The layer of soil below topsoil that contains more minerals and fewer organic materials

I. Substances that plants and animals need to grow

J. The layer under humus that provides nutrients and support for plant growth

Week Two

Day 5

Why is it important for farmers and gardeners to take care of the soil? What are some ways they can keep soil healthy and fertile?

Week Three

Fossils

Not all fossils are bones or shells. Trace fossils are footprints, burrows, or other evidence of an animal's activity. They tell us how ancient creatures moved and behaved.

Name: ______________________

Week Three

Dinosaur Fossils

Did you know that without fossils, we would not know much about dinosaurs? Dinosaurs became extinct long before humans existed, but fossils help scientists learn about how dinosaurs lived and what they looked like. It is fascinating to know that paleontologists have found dinosaur fossils on every continent on Earth, including Antarctica. **Paleontologists** are scientists who study fossils and the history of life. They study all kinds of fossils, not just dinosaur bones.

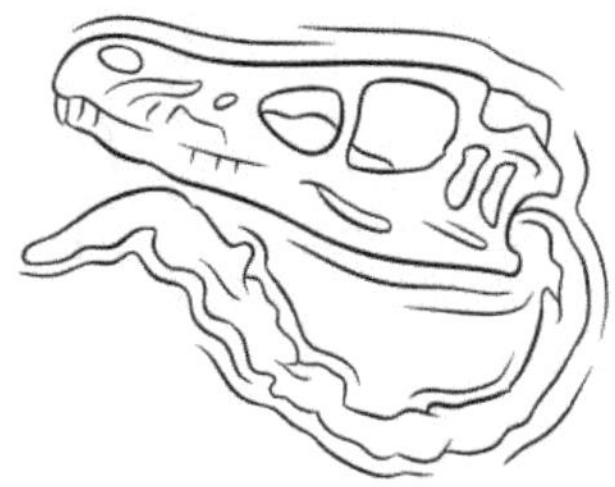

In ancient China, people once believed that dinosaur fossils were dragon bones. If you compare pictures of dragons and dinosaurs, you may notice some similarities. However, dinosaurs definitely did not breathe fire!

There are two main types of fossils: body fossils and trace fossils. **Body fossils** are the remains of the actual organism, such as bones, teeth, or shells. Mold fossils, cast fossils, and true-form fossils are all types of body fossils.

When a dinosaur dies and is buried quickly, sediment covers its remains. Over time, chemicals in the sediment cause changes to the remains. The bone slowly decays, and minerals replace the original material. This process is called **fossilization**. The end result is a heavy, rock-like copy of the original object, known as a **fossil**.

Trace fossils are clues left behind by an animal's activities, such as footprints, tracks, or burrows. For example, when a dinosaur walks through mud, the footprints can harden into rock and become trace fossils. Scientists use trace fossils to learn about how dinosaurs moved and behaved. Paleontologists certainly have interesting jobs, don't you think?

Types of Fossils

MOLD FOSSILS

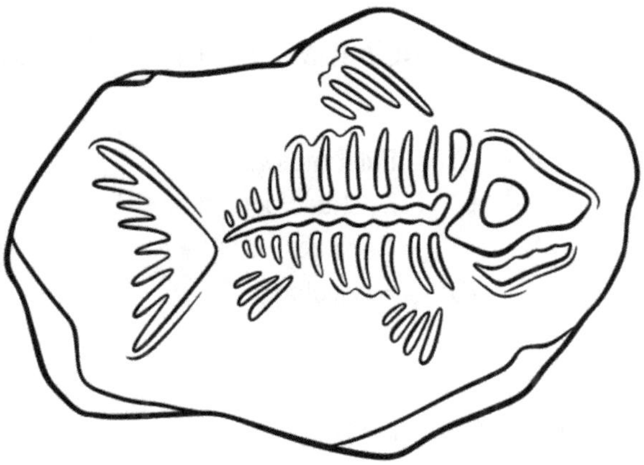

Mold fossils form when an organism is buried in sediment and then decomposes, leaving behind an impression or hollow space in the rock. This impression is a **negative image** of the organism's shape.

CAST FOSSILS

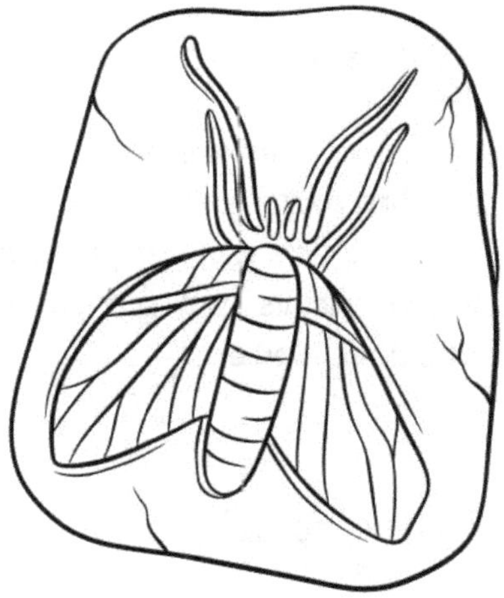

A cast fossil forms when a mold fossil is filled with minerals that harden into rock. Cast fossils create a **positive image** of the organism and show what the plant or animal looked like in three dimensions. Cast fossils are the opposite of mold fossils.

TRACE FOSSILS

Trace fossils are marks left behind by an animal's activity, such as footprints, nests, or burrows. For example, when an animal walks through mud and the mud hardens into rock, the footprints can become trace fossils. Scientists use trace fossils to learn about how animals moved and behaved.

TRUE-FORM FOSSILS

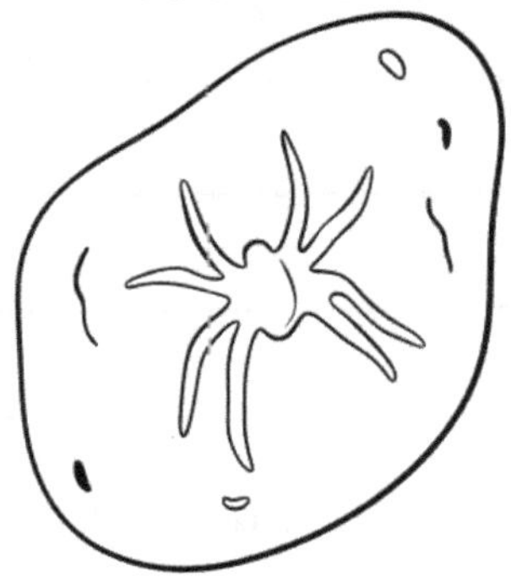

True-form fossils are the actual remains of an organism that have been preserved with little change. These fossils allow scientists to see what plants or animals were like long ago. True-form fossils can form when organisms are trapped in ice, tar, or amber, which helps preserve original features such as shape and color.

Week Three

Day 1

Name: ______________________________

Fossils

After reading the passage about fossils answer the following questions.

1. **How are fossils formed?**

   a) Instantly when an animal dies
   b) By heating rocks in a laboratory
   c) Over a long time as minerals replace the remains of plants or animals, turning them into hard rock
   d) Fossils don't even exist

2. **What is the study of fossils called?**

   a) Paleontology
   b) Archaeology
   c) Geology
   d) Fossilology

3. **What is a trace fossil?**

   a) A type of body fossil
   b) Marks left by animals, such as footprints, tracks, and burrows
   c) A fossil made from bones and teeth
   d) A fossil found only in Antarctica

4. **True or False: Dinosaurs were dragons.**

   a) True
   b) False

5. Explain the process of fossilization in your own words.

____

____

____

____

____

6. Describe the difference between body fossils and trace fossils.

____

____

____

____

____

Name: ______________________

Week Three

Day 2

Types of Fossils

Decide whether each description or picture is referring to a trace, mold, cast, or true-form fossil. Write your answers in the lines provided.

1.  Shows a negative image ______	**2.** The organism is preserved with little change ______	**3.** Marks that an animal left behind ______
4.  Animal trapped in amber ______	**5.** Mold filled in and then turned into rock ______	**6.** Shows an impression after the organism dissolves ______
7. An organism dies and minerals seep into the mold ______	**8.** Scientists use these fossils to study animal behavior ______	**9.** Scientists can make a cast fossil from it ______
10. 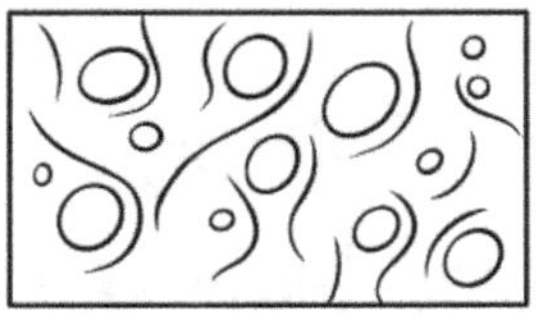 Shows the evidence of activity. Example: Burrows ______	**11.** 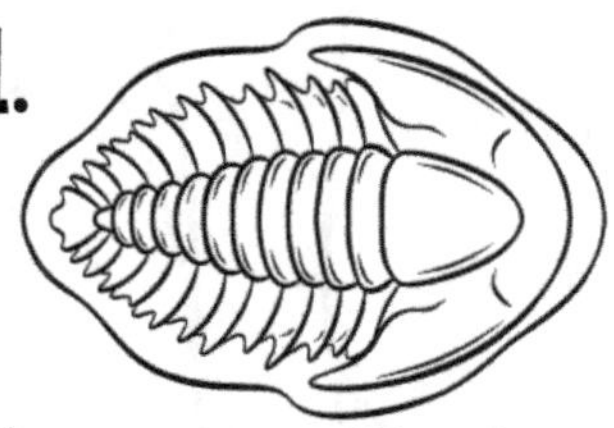 Shows a positive image ______	**12.** The actual plant or animal has been preserved ______

Name: ______________________

Process of Fossilization

Number the steps in the process of fossilization (from 1 to 6) to show the order in which they happen. Write each number in the correct circle.

◯ The sediment surrounding the skeleton thickens and begins to turn to stone.

◯ The fossil is exposed on Earth's surface because of wind and rain.

◯ Minerals enter the mold and fills the cavity, which forms a cast.

◯ The skeleton is buried by sediment.

◯ An animal dies and its soft parts rot away, leaving only the skeleton.

◯ The skeleton dissolves, leaving a mold behind.

Name: ______________________

Week Three

Day 4

Vocab Match

Match each word with its definition. Write the correct letter in the space provided.

1. Fossilization ________
2. Paleontologist ________
3. Decompose ________
4. Mold Fossils ________
5. Cast Fossils ________
6. Trace Fossils ________
7. True-Form Fossils ________
8. Body Fossils ________
9. Preserved ________
10. Sedimentary Rock ________

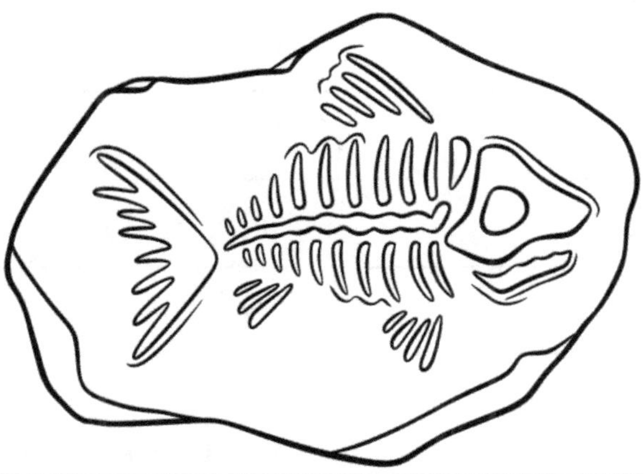

A. A fossil imprint made by the outside of an organism
B. Rock formed from layers of sediment that often contain fossils
C. The process by which the remains of living things are turned into rock
D. The remains of the actual organism, such as bones, teeth, or shells
E. Make or become rotten; decay
F. Marks left behind by an organism, such as footprints, tracks, or burrows
G. A scientist who studies fossils
H. Kept from decaying
I. The actual plant or animal; the whole organism has been preserved
J. A mold that was filled in and then turned into rock

Week Three

Day 5

Why is it important to study different types of fossils to get a complete picture of Earth's history? Explain.

Week Four

Landforms and Bodies of Water

Oceans cover about 71% of the Earth's surface and contain 97% of the Earth's water.

Name: ______________________

Week Four

Landforms

Landforms are natural features of the Earth's surface. They are all around us and come in many different forms, such as mountains, valleys, and canyons.

Landforms are created by natural processes such as rain, erosion, weathering, wind, earthquakes, and volcanic eruptions. These processes shape the land over time. For example, erosion caused by water and wind can wear down land and create landforms like valleys and canyons.

Mountains

A mountain is one of the highest landforms on Earth. It is usually wide at the bottom and narrow at the top, with steep rocky sides and a pointed top called a peak. The highest mountain range in the world is the Himalayas. Several mountains in a row are called a mountain range.

Hills

Hills are lower than mountains but are higher than their surrounding areas. They are usually covered with grass.

Valleys

Valleys are the low-lying areas between mountains or hills. Many people in ancient times settled in valleys because water runs down from the mountains, creating rich soil that is ideal for growing crops.

Plains

Plains are areas of flat land. The area of flat land along a sea or ocean is called a coastal plain. Most big cities are located on plains because it is easier to build houses, buildings, and roads.

Island

An island is a piece of land surrounded by water on all sides. Hawaii and Japan are examples of islands. A large group of islands close to each other forms an archipelago.

Name: ____________________________

Week Four

Landforms

Sand Dunes

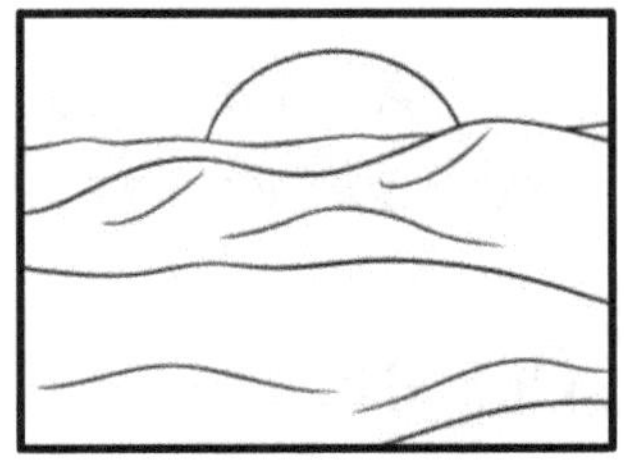

Sand dunes are huge hills of sand formed by wind in the desert or even the beach.

Peninsula

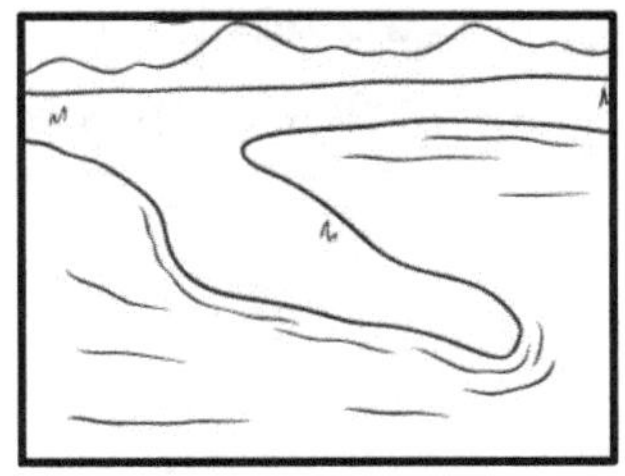

A peninsula is a piece of land that is surrounded by water from three sides. In the U.S., Florida is the most obvious peninsula

Canyon

A canyon has steep cliffs with a narrow valley running between them. It often has a river flowing through it. Canyons are created through erosion.

Delta

Deltas are wetlands that form as rivers empty their water and sediment into another body of water such as an ocean. As sediment builds up, new land is formed. This land is called a delta.

Plateau

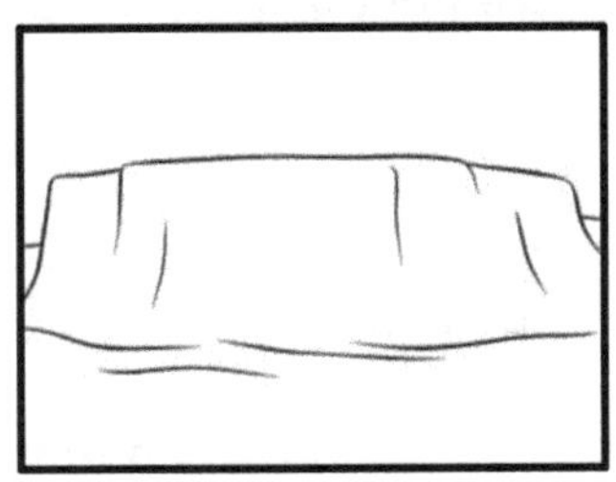

A plateau is an area of raised land that is flat on top. They are usually found in or near mountain ranges.

Mesa

A mesa is a medium size flat-topped mountain or hill with steep sides. They are usually found in dry places.

Even though Mesas and Plateaus both have flat tops and steep sides, they are different in size. A mesa is a smaller landform than a plateau.

Name: ______________________

Bodies of Water

A body of water is any large area that is covered by water. About three-fourths of the Earth's surface is covered with water in different forms and shapes.

Oceans

Oceans are large water bodies made up of salt water. We have five oceans in our world. The Pacific Ocean is the largest ocean covering one-third of the earth's surface.

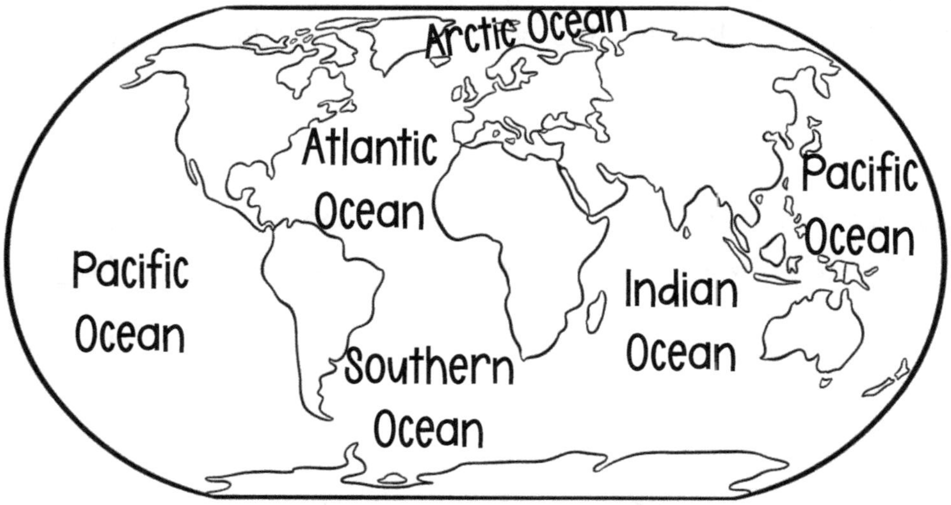

Seas

Seas are large bodies of water but are smaller than oceans. They are partly enclosed by land and connected to an ocean.

Lakes

A lake is a water body completely surrounded by land. Lakes can be salty or freshwater.

Pond

A pond is an area filled with water that is smaller than a lake. It can be natural or artificial.

Name: ______________________________

Week Four

Bodies of Water

Rivers

Rivers are large, flowing bodies of water that usually empty into a lake, sea, or ocean.

Gulf

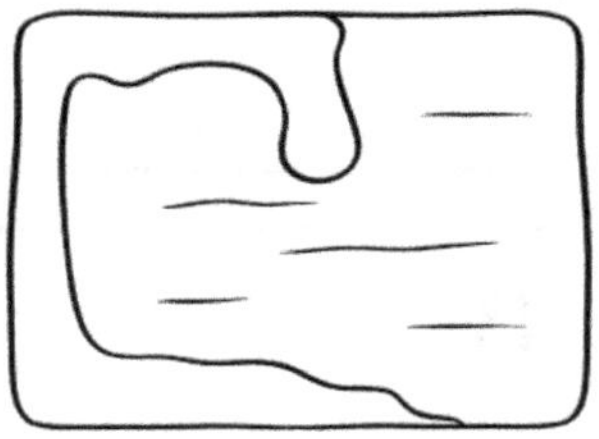

A gulf is a large area of an ocean or sea that is partially surrounded by land. For example: The Gulf of Mexico

Canal

Canals are waterways that are built by people and used for shipping and travel. A canal may also serve as a shortcut between two bodies of water. For example, People travel on canals in Venice Italy instead of using roads.

Glaciers

Glaciers are thick ice that covers a large area of land. They move very slowly. Around 10 percent of the world's land area is covered by glaciers.

Waterfall

A waterfall is water falling from a height. It forms when a river flows over the edge of hard rock and drops from a great height. Fun Fact: Angel Falls in Venezuela, South America, is the world's highest waterfall.

Name: ______________________

Week Four

Day 1

What Landform am I?

Read the passage about landforms. Then, identify the landform described in each statement below. Write the name of the landform in the provided space.

1. I am a piece of land surrounded by water on all sides. What am I? ______________	**2.** I am a huge hill of sand formed by wind. What am I? ______________	**3.** I am a piece of land surrounded by water on three sides. What am I? ______________
4. I am a low-lying land area between mountains or hills. What am I? ______________	**5.** I am one of the highest landforms on Earth with steep, pointed sides. What am I? ______________	**6.** I am an area of raised land that is flat on top. What am I? ______________
7. I am an area of steep cliffs and a narrow valley running in between. What am I? ______________	**8.** I am an area of flat land. What am I? ______________	**9.** I am lower than mountains but higher than my surroundings. What am I? ______________
10. I form as rivers empty their water and sediment into another body of water. What am I? ______________	**11.** I have a flat top and steep sides, but I am smaller than a plateau. What am I? ______________	**12.** A large group of islands close to each other forms me. What am I? ______________

Name: ______________________

Week Four

Day 2

What Body of Water am I?

Read the passage about bodies of water Then, identify the body of water described in each statement below. Write the name of the body of water in the provided space.

1. I am a large, flowing body of water that ends in an ocean or sea. ____________	**2.** I am a water body completely surrounded by land. I can be salty or freshwater. What am I? ____________	**3.** I am a waterway built by people and use me for shipping or travel. What am I? ____________
4. I am a large body of water but smaller than an ocean. I am partly enclosed by a land mass connected to an ocean. What am I? ____________	**5.** I am a natural or artificial area filled with water, but I am smaller than a lake. What am I? ____________	**6.** I am a large water body made up of salt water. What am I? ____________
7. I am a water falling from a height. What am I? ____________	**8.** I am a large area of an ocean and I am partially surrounded by land. What am I? ____________	**9.** I am a thick chunk of ice covering a large area of land. What am I? ____________

Week Four

Day 3

Name: ______________________________

Label the Landform

Word Bank	Peninsula	Canyon	Island	Valley	Hills	Plateau

1. 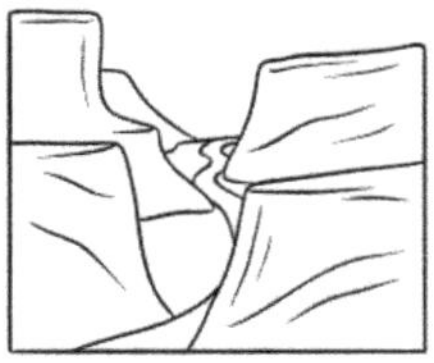 __________________

2.  __________________

3.  __________________

4.  __________________

5.  __________________

6. 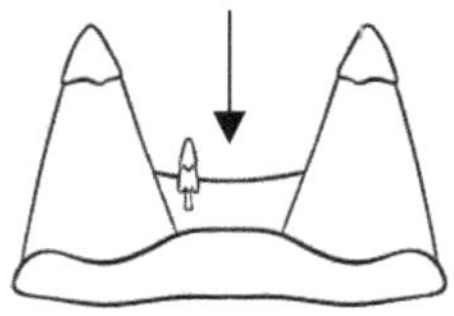 __________________

Label the Bodies of Water

Word Bank	Lake	Gulf	Canal	Waterfall	Ocean	River

7. 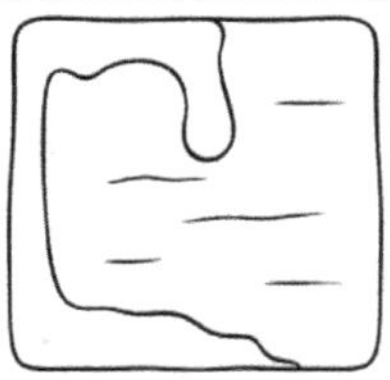 __________________

8. 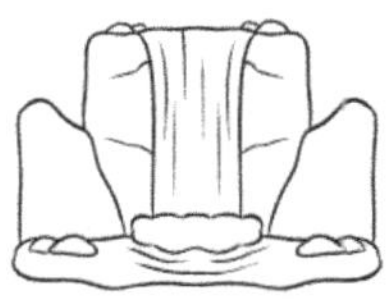 __________________

9. 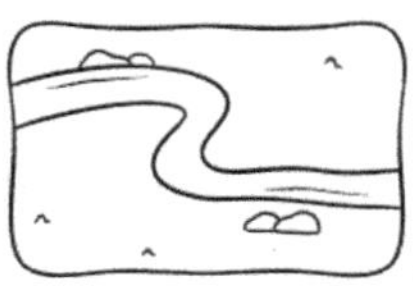 __________________

10. 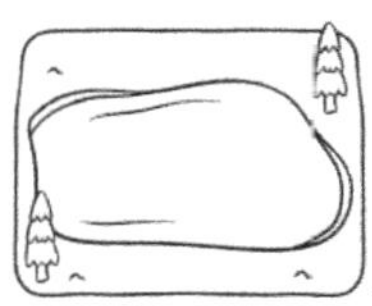 __________________

11. 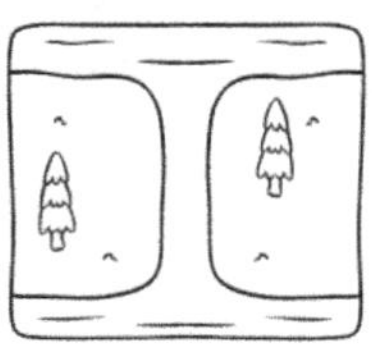 __________________

12.  __________________

Week Four

Day 4

Name: ______________________________

Landforms and Bodies of Water

After learning about landforms and bodies of water, answer the questions below.

1. **Which of the following is a type of landform?**

   a) River
   b) Peninsula
   c) Ocean
   d) Lake

2. **What is an archipelago?**

   a) A large, flat area of land
   b) A single mountain
   c) A deep valley
   d) A group of islands

3. **What landform is created by the deposition of sediment at the mouth of a river?**

   a) Delta
   b) Plateau
   c) Canyon
   d) Plain

4. **Why are most big cities located in plains?**

   a) Plains are closer to the sun, making them warmer
   b) Plains have less wildlife, making them safer for people.
   c) It is easier to build houses, buildings, and roads.
   d) Because plains are surrounded by water on all sides, and people need water to live.

5. **Describe the process of how a waterfall forms and provide an example of a famous waterfall.**

____

____

____

____

____

____

6. **Explain the similarities and differences between a plateau and a mesa.**

____

____

____

____

____

Week Four

Day 5

What are the benefits and challenges of living near a lake or a pond?

Week Five

The Water Cycle

A single water molecule can take thousands of years to complete a journey through the water cycle, moving between the ocean, atmosphere, and land.

Week Five

Name: ______________________________

The Water Cycle

Did you know that the Earth has been recycling water for over 4 billion years? As far as we know, no new water has formed. This means there is a very high chance that the water you drink today is the same water that dinosaurs drank millions of years ago!

Water is all around us. About 71% of Earth is covered in water. This water is constantly moving through what we call the **water cycle**. The water cycle never stops and does not really have a beginning or an end. It is made up of four main stages: evaporation, condensation, precipitation, and collection.

Evaporation is the process by which water changes from a liquid into a gas called water vapor. The sun heats water in rivers, lakes, and oceans, causing it to turn into water vapor and rise into the air. When this happens, the water is no longer a liquid. It becomes a gas.

Condensation is the next stage of the water cycle. As water vapor rises higher into the atmosphere, the cooler air causes it to change back into a liquid. These tiny water droplets come together to form clouds.

When clouds become too full of water, the water falls back to Earth as **precipitation**. Precipitation can be rain, snow, sleet, or hail. Some precipitation evaporates again, while some flows over the land into creeks, rivers, and streams. This movement of water is called **runoff**. When water soaks into the ground through soil and rocks, it is called **infiltration**.

Collection is the stage after precipitation when water gathers in oceans, lakes, rivers, and other bodies of water. Some water also remains on land and is absorbed by plant roots. The water cycle then begins again as water evaporates back into the atmosphere.

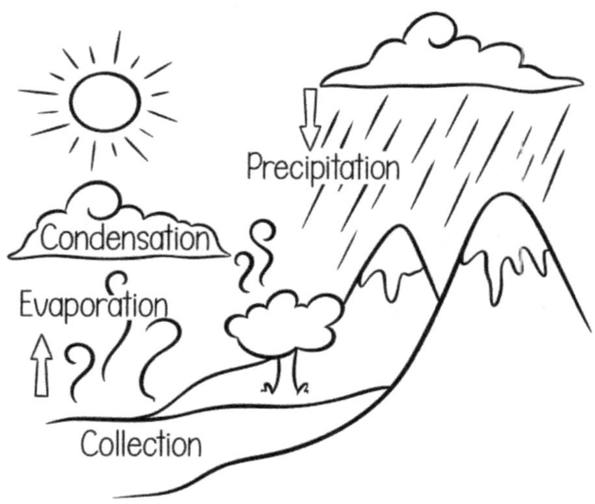

Name: ______________________

Week Five

Day 1

The Water Cycle

1. **What is the process by which water changes from a liquid to a gas?**
   a) Evaporation
   b) Condensation
   c) Precipitation
   d) Infiltration

2. **Which of the following describes the water cycle?**
   a) The movement of water from oceans to lakes
   b) The continuous cycle of water on Earth
   c) The process of water freezing into ice
   d) None of the above

3. **What happens during the process of condensation?**
   a) Water turns into ice
   b) Water droplets form and come together to form clouds
   c) Water falls from the sky as rain, snow, or hail
   d) Water runs off the mountain

4. **What are examples of forms of precipitation?**
   a) Rivers and lakes
   b) Clouds and fog
   c) Rain, snow, and sleet
   d) Condensation

5. **Why is the water cycle important for life on Earth?**

____

____

____

____

____

6. **What role do clouds play in the water cycle?**

____

____

____

____

____

Name: ______________________

The Water Cycle

Label the water cycle and then fill in the missing words in the sentences below.

Word Bank: Condensation, Runoff, Infiltration, Collection, Transpiration, Evaporation, Precipitation

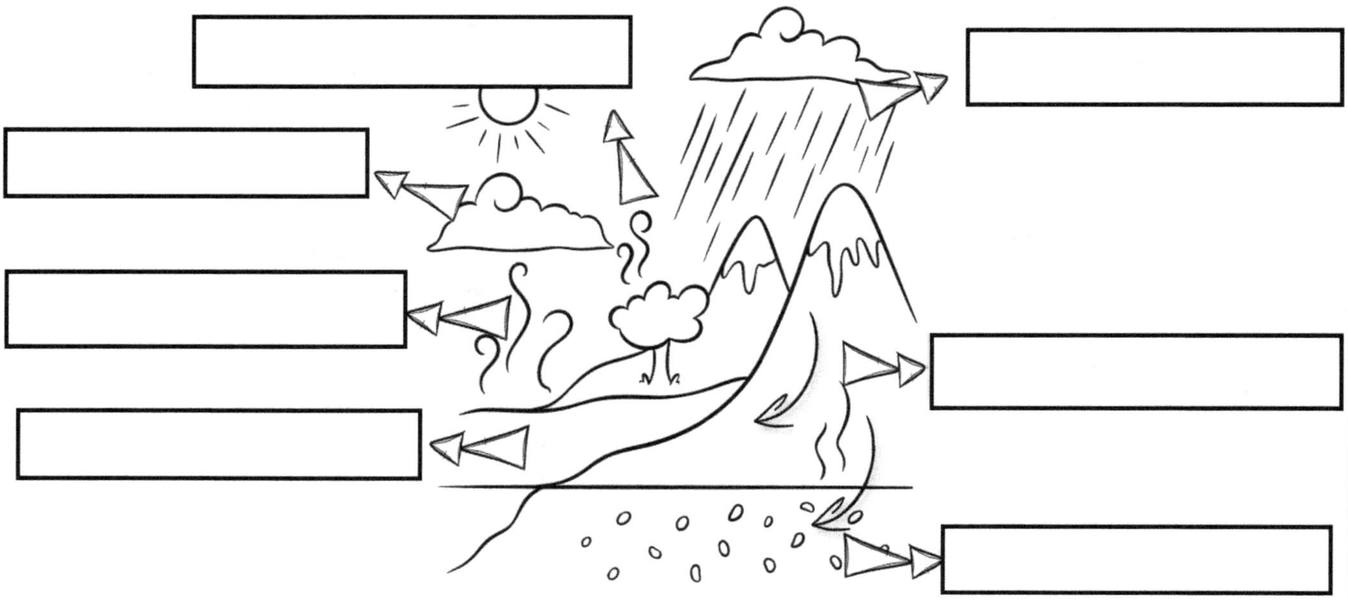

Word Bank: Rocks, Precipitation, Evaporate, Sun, Gas, Infiltration

1. There would not be a water cycle without the ______________.
2. Evaporation is the process where liquid becomes ______________.
3. When water soaks into the ground through soil and ______________ we call it ______________.
4. Runoff is ______________ that did not get absorbed into the soil or did not ______________ into the atmosphere.

THE WATER CYCLE

Week Five

Day 3

Name: ______________________________

Look at each picture and identify the water cycle process shown. Write the correct term from the word bank on the lines provided. Number 5 is already labeled. In the last box, draw and label your own example of a water cycle process.

Word Bank

Evaporation   Transpiration   Condensation   Precipitation   Infiltration   Runoff

1. ______________________________

2. ______________________________

3. ______________________________

4. ______________________________

5. Collection

Water gradually gathers in large bodes

6. ______________________________

7. ______________________________

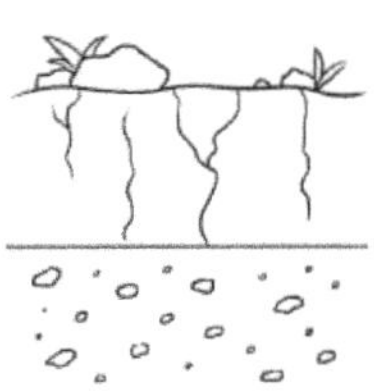

Draw Your Own Water Cycle Example

Week Five

Day 4

Name: ____________________

Vocab Match

Match each word with its definition. Write the correct letter in the space provided.

1. Water Cycle ________
2. Infiltration ________
3. Evaporation ________
4. Precipitation ________
5. Hydrology ________
6. Transpiration ________
7. Water Vapor ________
8. Condensation ________
9. Collection ________
10. Runoff ________

A. The process where liquid turns into gas

B. Rain, snow, sleet, or hail that falls from clouds to the ground

C. Precipitation that does not soak into the soil and flows over land to rivers, lakes, or oceans

D. Process where water soaks into the ground through soil and rocks

E. The release of water vapor from plant leaves into the air

F. The continuous movement and changes of water as it travels through Earth

G. The process where gas turns into liquid

H. Process in which water gradually gathers in large bodies such as oceans, lakes, and rivers

I. Invisible water in the form of a gas

J. The study of water on Earth, including water in the air, on land, and underground

Week Five

Day 5

How does the water cycle impact ecosystems and the environment?

Week Six

Weather and Climate

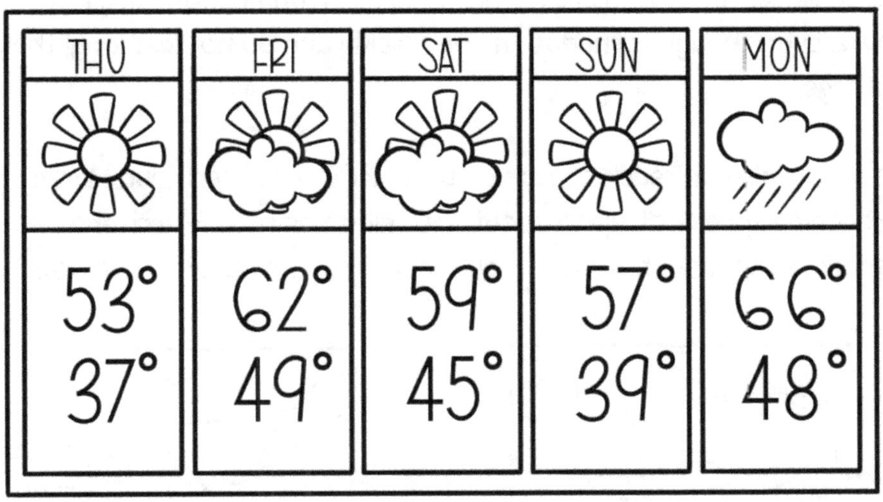

Weather describes the short-term conditions of the atmosphere at a specific time and place, like rain, snow, or sunshine. Climate is the average weather in a place over a long period.

Name: ______________________________

Week Six

Day 1

Tornadoes Vs. Hurricanes

Tornadoes and hurricanes are powerful storms with strong spinning winds that can cause serious damage. Both storms can be dangerous, but they are different in many ways.

Tornadoes usually form over land, while hurricanes form over warm ocean waters in tropical areas. One of the main differences between tornadoes and hurricanes is their size and how long they last. Tornadoes are usually only a few hundred feet wide, while hurricanes can stretch across hundreds of miles. Hurricanes can last for several days or even weeks, but tornadoes usually last only a short time, often less than an hour.

Tornadoes happen more often than hurricanes. Each year, about 800 to 1,000 tornadoes occur in the United States. In comparison, only about 10 tropical storms develop into hurricanes in the Atlantic Ocean each year.

Hurricanes can often be predicted days in advance, giving people time to prepare or evacuate to a safer place. Tornadoes are harder to predict, and people may have only 15 to 30 minutes to prepare. Even though tornadoes move quickly and do not last long, they can still cause a great deal of damage.

Winds in a tornado can reach speeds of up to 300 miles per hour. Hurricane winds are usually slower, reaching speeds of up to about 180 miles per hour. Hurricanes gain strength while moving over warm ocean water, but once they reach land, they begin to weaken.

TORNADO

HURRICANE

Week Six

Day 1

Name: ______________________________

Tornadoes Vs. Hurricanes

		TORNADOES	HURRICANES
1.	How do they form?		
2.	How big are they?		
3.	How long do they last?		
4.	How strong are the winds?		
5.	How common are they?		

6. Do you usually have enough time to plan and evacuate if a hurricane is coming to your area? Explain your answer.

7. About how much time do people usually have to prepare for a tornado?

8. What happens to a hurricane once it hits land?

Name: ____________________

Weather Instruments

Weather instruments are special tools that scientists, called meteorologists, use to study and predict the weather. These instruments help meteorologists collect information about the atmosphere. They study this information to make forecasts, which are predictions about what the weather will be like in the next few hours, days, or even weeks. Below are some common weather instruments and how they work.

THERMOMETER

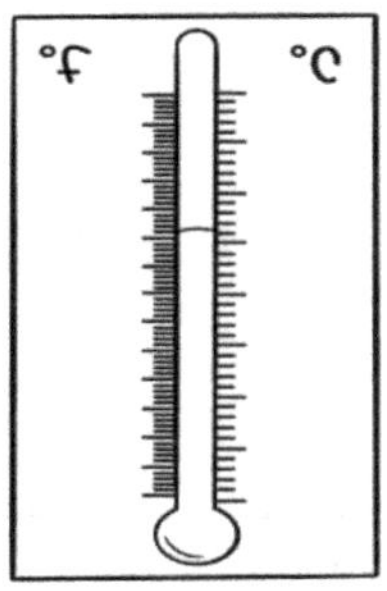

A thermometer measures the temperature of the air. It tells us how hot or cold it is outside. Some thermometers are made of glass and contain a liquid inside. When the air gets warmer, the liquid rises. When the air gets cooler, the liquid falls.

BAROMETER

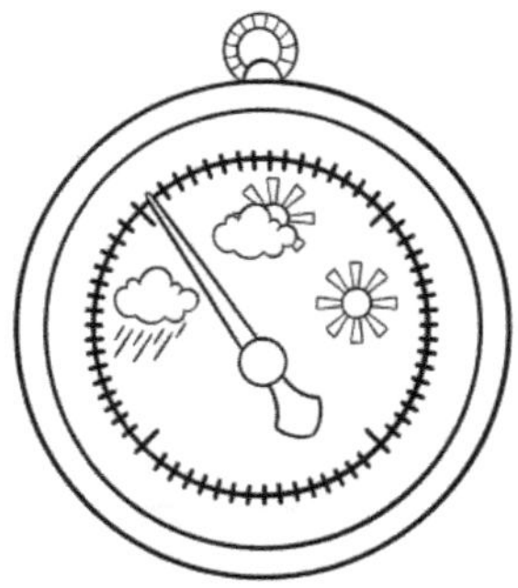

A barometer measures air pressure. Air pressure can tell us a lot about the weather. High air pressure usually means clear and calm weather, while low air pressure can mean stormy weather is coming.

ANEMOMETER

An anemometer measures wind speed. It has cups or blades that spin faster when the wind blows harder. Knowing how fast the wind is blowing helps scientists understand weather patterns and predict storms.

RAIN GAUGE

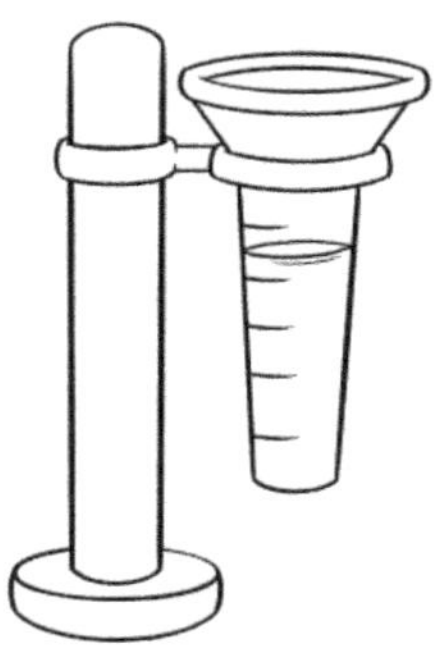

A rain gauge measures how much rain has fallen. It is usually a clear container with measurement markings on the side. By checking the rain gauge, we can see how much rain fell during a storm.

WIND VANE

A wind vane, also called a weather vane, shows the direction the wind is coming from. It usually has an arrow that points into the wind. This information helps scientists predict how weather systems will move.

HYGROMETER

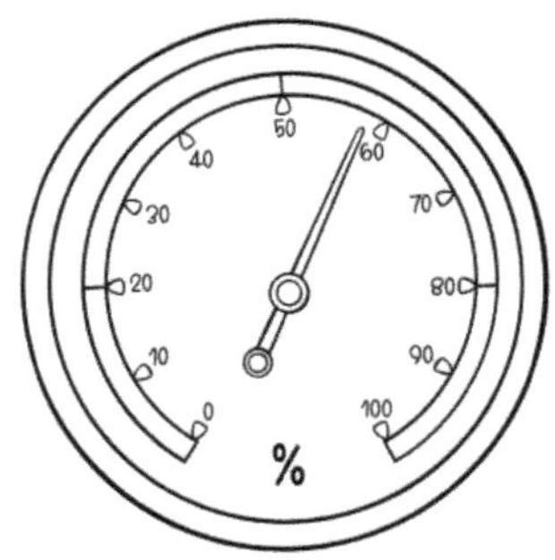

A hygrometer measures humidity, which is the amount of moisture in the air. High humidity can make the air feel sticky and can lead to rain. Low humidity makes the air feel dry.

Week Six

Day 2

Name: ______________________________

Weather Instruments

1. Rain Gauge __________
2. Thermometer __________
3. Barometer __________
4. Wind vane __________
5. Anemometer __________
6. Hygrometer __________

(A) Measures wind speed ______________________

(B) Measures air pressure ______________________

(C) Measures humidity ______________________

(D) Measures how much rain has fallen ______________________

(E) Shows the direction the wind is coming from ______________________

(F) Measures the temperature of the air

Write the name of the weather instrument under each picture.

7. 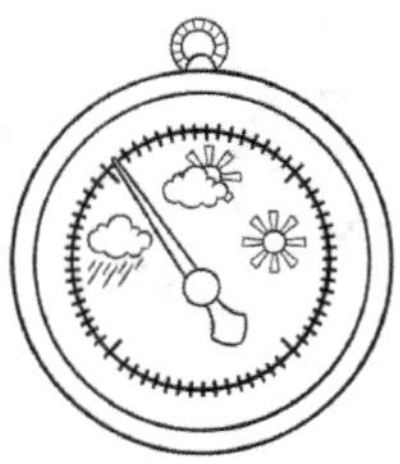 ____________	8. 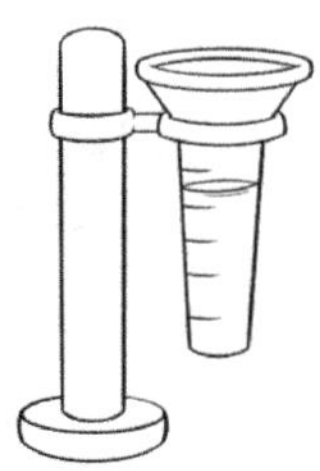 ____________	9.  ____________
10.  ____________	11.  ____________	12. 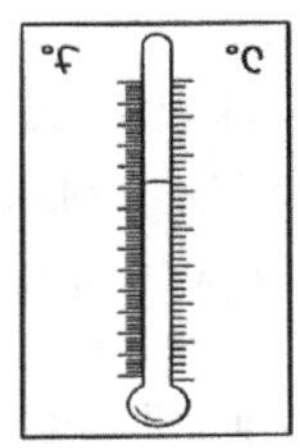 ____________

Name: ______________________________

Week Six

Day 3

Weather Vs. Climate

Weather describes the conditions outside at a specific time and place. It can change quickly, from sunny to rainy or from warm to cold in just a few hours. For example, it may be sunny in the afternoon and rainy at night. Weather includes factors such as temperature, wind, humidity, and precipitation, including rain, snow, sleet, and hail.

Climate, on the other hand, is the average weather in a place over a long period of time, usually 30 years or more. Climate tells us what kind of weather to expect during different seasons. For example, a desert has a dry climate, while a rainforest has a wet climate. Understanding climate helps us know the usual weather patterns in different parts of the world.

Both weather and climate are important because they affect how we live, what we wear, and even what we eat. They also help scientists understand how Earth is changing over time.

Take a look at the general climate zones of the world below.

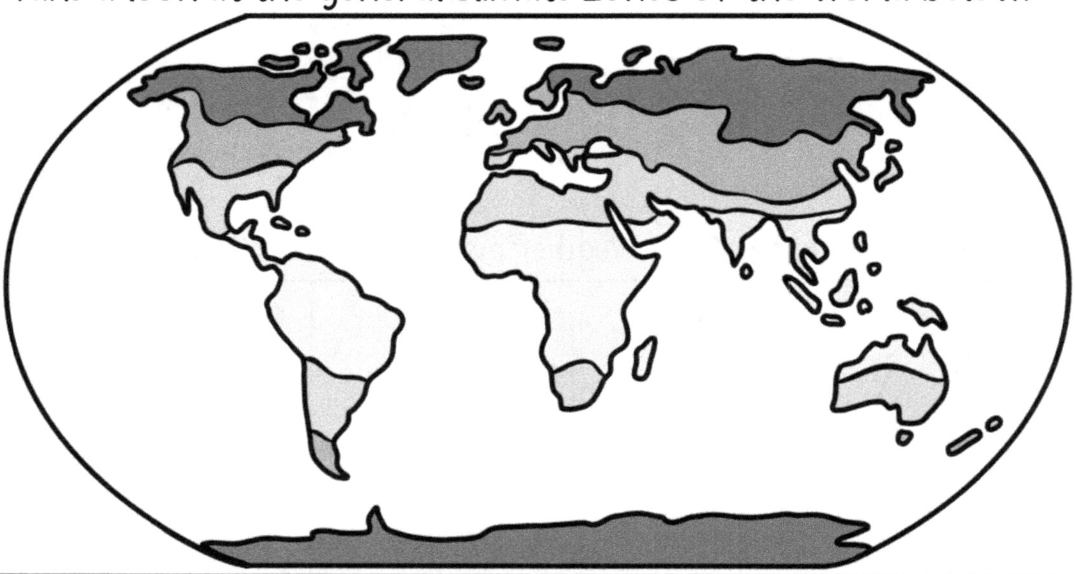

Polar and Subpolar	Temperate	Subtropical	Tropical
Near the North and South poles. Very cold all year round, with long, harsh winters and short, cool summers. There is often ice and snow covering the ground.	Between the polar and tropical zones. Moderate temperatures with four seasons: winter, spring, summer, and fall. Winters are cold and simmers are warm.	Between the temperate and tropical zones. Generally warm, with hot summers and mild to cool winters. There can be more variation in rainfall, with wet and dry seasons.	Near the equator. Very warm and humid all year round. There is a lot of rainfall, especially in tropical rainforests. This zone does not experience four distinct seasons like temperate zones.

Name: ______________________

Week Six

Day 3

Weather Vs. Climate

1. What is weather?

a) The long-term average of atmospheric conditions
b) The current conditions outside at a specific time and place
c) The study of stars and planets
d) A type of ecosystem

2. Which of the following is an example of climate?

a) A thunderstorm on a summer afternoon
b) A snowstorm that lasts for one day
c) The average temperature and rainfall in a rainforest over 30 years
d) A windy day in the spring

3. What does precipitation include?

a) Sunlight and wind
b) Rain, snow, sleet, and hail
c) Clouds and fog
d) Temperature and humidity

4. How is climate different from weather?

a) Climate changes daily, while weather remains the same
b) Climate describes long-term patterns, while weather describes short-term conditions
c) Weather includes only temperature, while climate includes only precipitation
d) Climate is only about wind, while weather is only about rain

---

5. Which climate zone is near the equator and has very warm and humid weather all year round?

a) Polar
b) Subpolar
c) Temperate
d) Tropical

6. What is a characteristic of the temperate zone?

a) Moderate temperatures with four distinct seasons
b) Extremely cold temperatures all year
c) Hot and humid weather with no distinct seasons
d) Very dry with little rainfall

7. Which climate zone has long, harsh winters and is found near the North and South Poles?

a) Tropical
b) Temperate
c) Polar
d) Subtropical

8. Where would you find a subtropical climate zone?

a) Near the equator
b) Between the polar and temperate zones
c) Just below the polar zones
d) Between the temperate and tropical zones

Name: ______________________________

Week Six

Day 4

Air Pressure and Weather Fronts

Meteorologists study high and low air pressure systems to help predict the weather. Differences in air pressure cause wind to form. **Air pressure** is the weight of the air pressing down on the Earth. When air pressure is high, we usually experience clear and sunny weather. When air pressure is low, clouds, rain, or storms are more likely to occur.

Low Pressure System: Brings storms and strong winds. It attracts warm, moist air, which can lead to cloudy, rainy, or stormy weather.

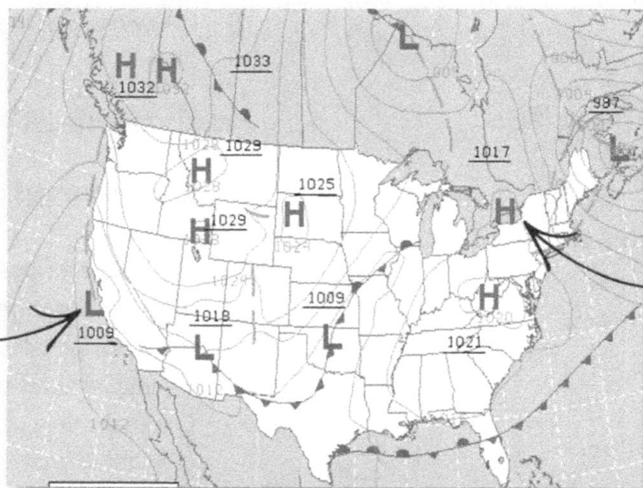

High Pressure System: Brings good weather and sunny skies. It attracts cool, dry air and often pushes clouds away.

The boundary between high and low air pressure systems is called a **weather front**. Weather fronts form when different types of air meet. When two air masses meet, storms often begin to form. There are four main types of weather fronts: cold fronts, warm fronts, stationary fronts, and occluded fronts.

Name: ______________________

Week Six

Day 4

Weather Symbols

Read about each symbol and **color each symbol** with its correct color.

COLD FRONT

On weather maps, a cold front symbol is shown as a **blue line** with **blue triangles** pointing in the direction the cold air is moving.

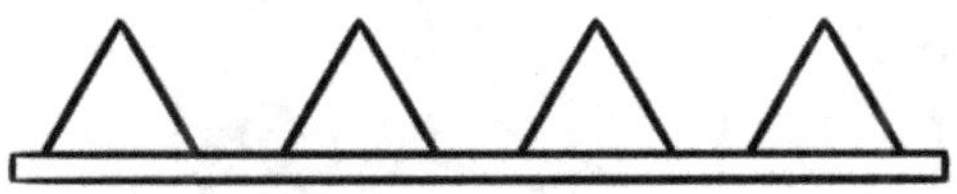

A cold front forms when a cold air mass replaces a warm air mass. Temperatures usually drop, and it can cause a narrow line of storms.

WARM FRONT

On weather maps, a warm front is shown as a **red line** with **red semicircles** pointing in the direction the warm air is moving.

A warm front forms when warm air moves into a colder area. It usually brings gradual warming, steady rain, or light snow.

STATIONARY FRONT

On weather maps, a stationary front is shown as a line with alternating **blue** and **red** sections, with **blue triangles** on the blue parts and **red semicircles** on the red parts, pointing in opposite directions.

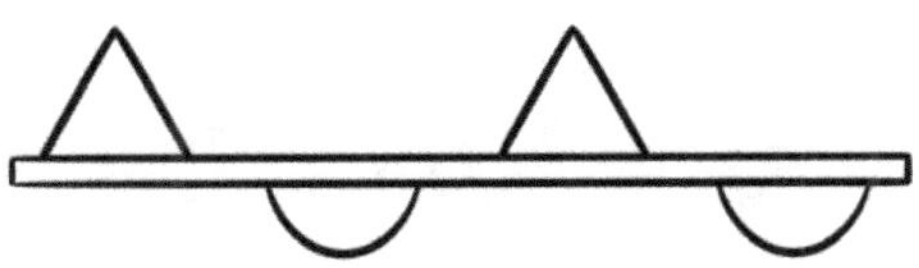

A stationary front stays in one place for a long time and can bring long periods of rain or cloudy weather.

OCCLUDED FRONT

On weather maps, an occluded front is shown as a **purple line** with both **purple** triangles and semicircles.

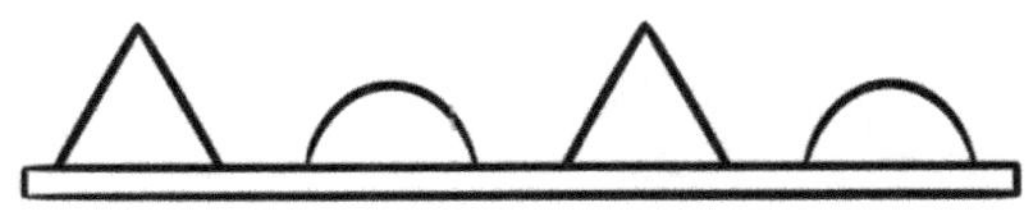

An occluded front forms when a cold front catches up to a warm front.

HIGH AIR PRESSURE

On weather maps, high air pressure is shown with a **blue** "H."

High air pressure occurs when air presses down on the Earth's surface more than usual. It usually brings clear, calm weather because it pushes clouds away.

LOW AIR PRESSURE

On weather maps, low air pressure is shown with a **red** "L."

Low air pressure occurs when air is not pressing down on the Earth's surface as much. It often brings cloudy, rainy, or stormy weather because it allows clouds to form and grow.

Name: ______________________________

Week Six

Day 4

Let's Create a Weather Map!

Read the directions and draw and color the symbols on the map to represent the information. Use the correct colors and symbols shown on the previous page when drawing each weather feature.

1. Draw a low pressure system over California.
2. Draw a cold front moving through Texas, Louisiana, and Mississippi.
3. Draw a high pressure system over Florida.
4. Draw a warm front moving north through Ohio.
5. Draw an occluded front in New York.
6. Draw a low pressure system over Colorado.
7. Draw a stationary front across Virginia.
8. Draw a high pressure system over Oregon.

Name: ______________________

Week Six

Day 5

1. How do weather patterns in your area affect the activities you can do throughout the year? Think about the different seasons and types of weather where you live. How do they impact outdoor activities, sports, and daily routines?

2. How might a sudden change in climate affect the environment and the people living in it? Imagine your area suddenly became much hotter, colder, wetter, or drier. How would this change affect plants, animals, and humans?

Week Seven

Volcanoes

Volcanic eruptions can be explosive, sending ash and gas high into the sky, or they can be gentle, where lava flows steadily out of the volcano.

Name: ______________________

Week Seven

Earth's Powerful Forces: Volcanoes

Exploding volcanoes show us the power of nature and what it is capable of. Did you know that a volcano is an opening in Earth's surface? Magma, hot gases, ash, and rock fragments escape from inside Earth through this opening during an eruption. When magma reaches the Earth's surface, it is called **lava**.

There are different types and stages of a volcano.

TYPES OF VOLCANOES

SHIELD VOLCANOES

A shield volcano has a wide base and a large bowl-shaped opening at the top.. It is made by repeated eruptions producing lava flows. The Hawaiian Islands have shield volcanoes.

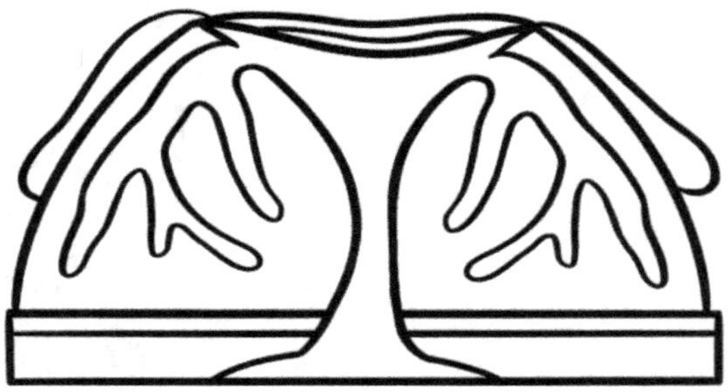

COMPOSITE VOLCANOES (ALSO KNOWN AS STRATOVOLCANOES)

Composite volcanoes are the tallest and the steepest volcanoes. They are made of many layers of volcanic rock formed from lava, ash, and rock debris. Mount Fuji in Japan is an example of a composite volcano.

CINDER CONE VOLCANOES

Cinder cone volcanoes are one of the most common types of volcanoes. They are the smallest types of volcanoes and they have steep sides. They form from volcanic rocks called cinders that pile up around the vent. An example is Paricutin in Mexico.

Name: ______________________

Week Seven

STAGES OF VOLCANOES

ACTIVE

- Currently erupting or has recently erupted
- Can show signs that it may erupt again, such as smoke, ash, or small tremors.
- Examples: Mount St. Helens (USA), Mount Etna (Italy)

DORMANT

- Has not erupted in a long time, but it shows signs that it will erupt again in the future
- Sometimes called a "sleeping" volcano.
- Examples: Mauna Kea of Hawaii (USA), Mount Fuji (Japan)

EXTINCT

- Has not erupted for thousands of years and is unlikely to erupt again
- Considered "dead" because it no longer has a magma supply.
- Examples: Huascarán (Peru), Mount Buninyong (Australia)

Name: ______________________________

Week Seven

Day 1

Volcanoes

1. **Which type of volcano is the tallest and is made from layers of lava and ash?**

   a) Shield Volcano
   b) Composite Volcano
   c) Cinder Cone Volcano
   d) Lava Volcano

2. **What is a characteristic of a cinder cone volcano?**

   a) It has gentle slopes and is very wide
   b) It is tall and made of layers of lava and ash
   c) It is small with steep sides made of volcanic rocks called cinders
   d) It is flat and made from slow-moving rocks called cinders

3. **Which stage of a volcano is currently erupting or has erupted recently?**

   a) Active
   b) Dormant
   c) Extinct
   d) Sleeping

4. **What is a characteristic of an extinct volcano?**

   a) It is currently erupting
   b) It has the potential to erupt in the future
   c) It has not erupted for thousands of years and is unlikely to erupt again
   d) It shows signs of activity like smoke and ash

5. **What is the difference between magma and lava?**

____

____

____

6. **Describe the difference between a shield volcano and a composite volcano.**

____

____

____

____

7. **Explain what it means for a volcano to be dormant.**

____

____

____

____

Week Seven

Day 2

Name: ____________________________

Vocab Match

Match each word with its definition. Write the correct letter in the space provided.

1. Magma ________
2. Lava ________
3. Eruption ________
4. Dormant Volcano ________
5. Active Volcano ________
6. Extinct Volcano ________
7. Molten ________
8. Volcanic Ash ________
9. Vent ________
10. Crater ________

A. A volcano that has not erupted recently but is expected to erupt again

B. An opening at the top or side of a volcano where lava erupts

C. Becoming liquid because of heat

D. Molten rock located deep below the Earth's surface

E. The bowl-shaped area around the opening of a volcano

F. A volcano that erupted thousands of years ago and is unlikely to erupt again

G. The sudden release of lava, ash, and gases from a volcano

H. Tiny pieces of rock, minerals, and volcanic glass released into the air during an eruption

I. Magma that reaches the Earth's surface

J. A volcano that is erupting now or has erupted recently

Name: ______________________

Week Seven

Day 3

If you were a volcano...

Pretend you are a volcano. Answer the questions and draw what your volcano would look like in the space provided.

VOLCANO NAME:

LOCATION:

HOW I LOOK LIKE:

Did you cause destruction? Explain.

Are you a composite, cinder cone, or shield volcano? How do you know that?

(circle one)
I am...

ACTIVE

DORMANT

EXTINCT

Week Seven

Day 4

Name: ____________________

Parts of a Volcano

Use the word bank to label the parts of a volcano.

Word Bank: Magma, Crater, Side Vent, Lava, Ash Cloud, Main Vent

VOLCANOES

Week Seven

Day 5

List and explain the dangers that volcanic eruptions can pose to people and the environment. If possible, watch a short video of a volcanic eruption before writing.

Name: ______________________________

THE VOLCANO Experiment

Materials

- ✓ 2-liter soda bottle
- ✓ ¼ cup Warm water
- ✓ 5 drops of food coloring
- ✓ 1 tsp of dish Soap
- ✓ 2 tbsp of baking Soda
- ✓ ½ cup of vinegar

Directions

1. Get an empty 2-liter soda bottle and place it outside or on a tray. Carefully pour ¼ cup of warm water into the bottle.
2. Add 5 drops of food coloring and 1 teaspoon of dish soap to the bottle. Gently swirl the bottle to mix.
3. Add 2 tablespoons of baking soda to the bottle.
4. Quickly and carefully pour ½ cup of vinegar into the bottle. Step back and watch the eruption!

Describe what happened.

Draw what happened.

Why do you think this eruption happened?

If you add more vinegar, how will it change the eruption? Explain.

If you add more baking soda, how will it change the eruption? Explain.

Week Eight

Plate Tectonics

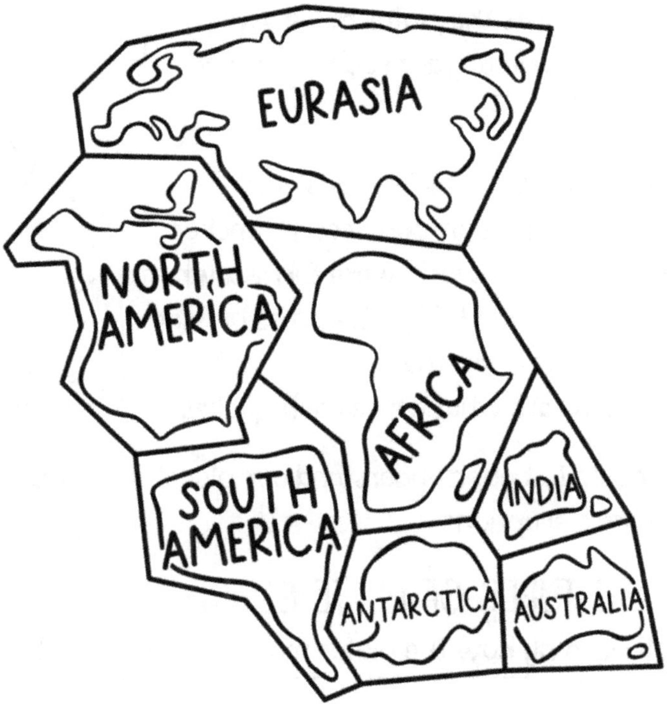

Tectonic plates are always moving, but very slowly. They move at a rate of about 1 to 2 inches per year, about as fast as your fingernails grow!

Name: ______________________________

Week Eight

Exploring Plate Tectonics

The Earth's surface is made up of large, moving pieces called tectonic plates. Think of them like a giant jigsaw puzzle that fits together to form the Earth's surface. There are seven major plates and many smaller plates. The plates are made of rock and are constantly moving around the planet. You usually cannot notice their movement because they only move about 1 to 2 inches per year, about as fast as your fingernails grow.

Tectonic plates are pieces of the Earth's **lithosphere**. The lithosphere is the outermost layer of the Earth made up of the crust and the upper part of the mantle. The plates float on top of a hotter and more fluid layer called the **asthenosphere**. Because the asthenosphere is more fluid, it allows the plates to move in different directions. Sometimes the plates move away from each other, sometimes they move toward each other, and sometimes they slide past each other.

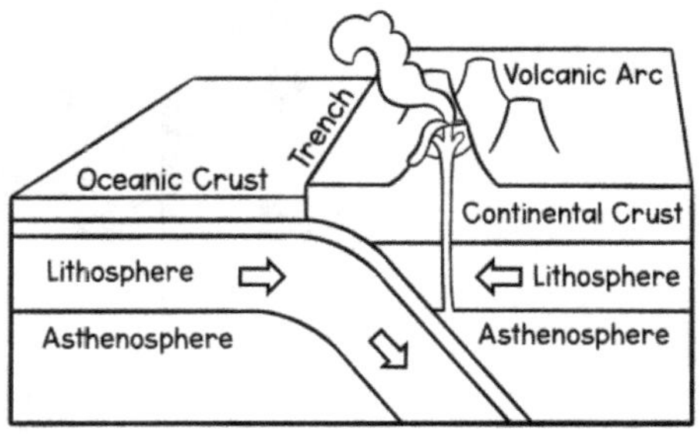

The theory of plate tectonics explains the causes of earthquakes, volcanoes, mountain formation, and ocean trenches. For example, the tallest mountains in the world were formed when two plates collided and pushed up the land. If plates press or slide against each other too much, the force can cause vibrations that shake the ground. These vibrations are what cause earthquakes.

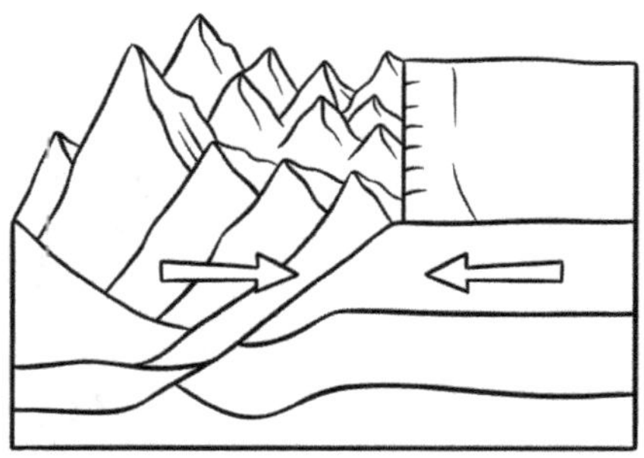

Scientists study plate movements to better understand our planet and to help predict natural events such as earthquakes and volcanic eruptions.

3 TYPES OF PLATE BOUNDARIES

(characterized by how the plates move relative to each other)

TRANSFORM	DIVERGENT	CONVERGENT
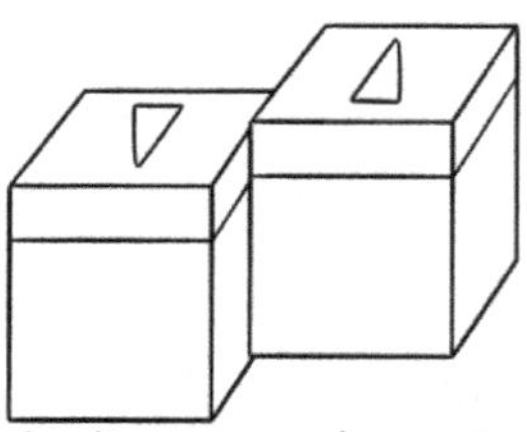	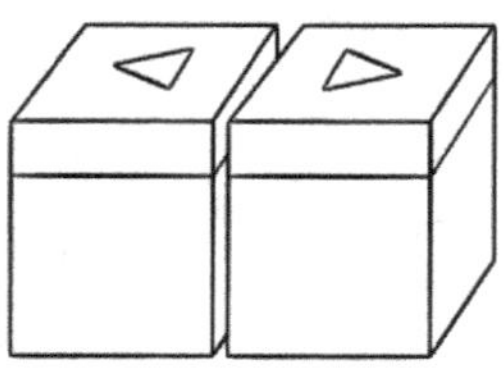	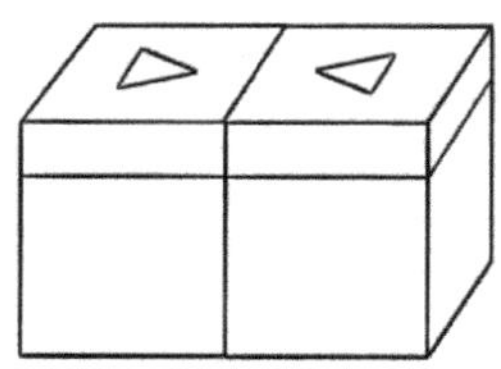
- A fault zone where two plates slide past each other - Earthquakes often occur in these areas.	- Two plates move away from one another - Divergent boundaries within continents produce rift valleys	- Two plates push together - Areas of mountain formations, volcanoes, and high earthquake activity

Week Eight

Name: ______________________________

THE MAJOR TECTONIC PLATES OF EARTH

African Plate

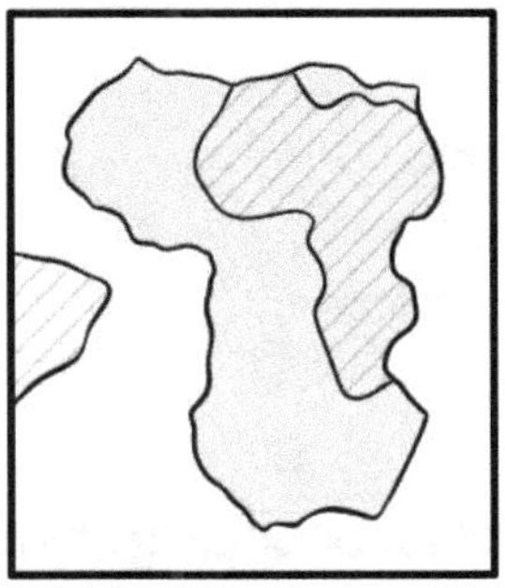

Pacific Plate

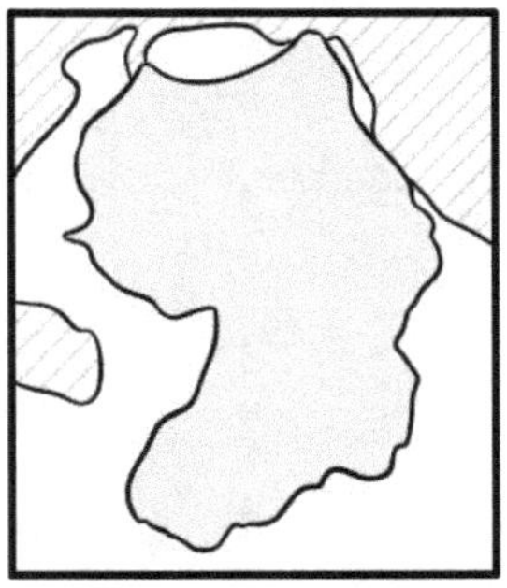

North American Plate

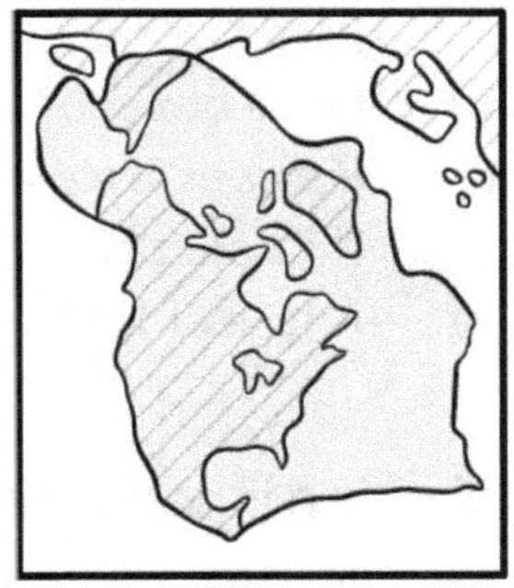

Antarctic Plate

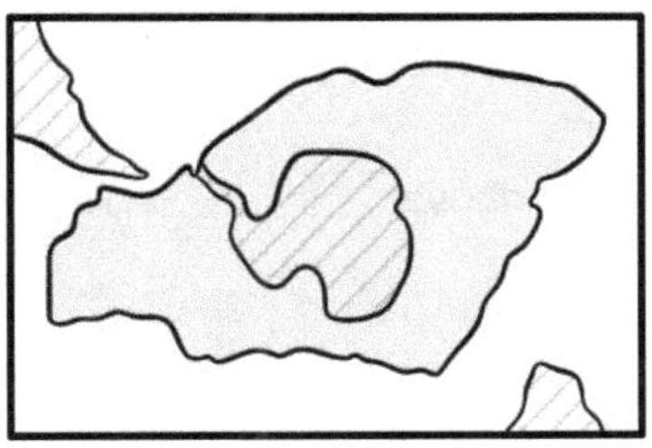

Australian Plate

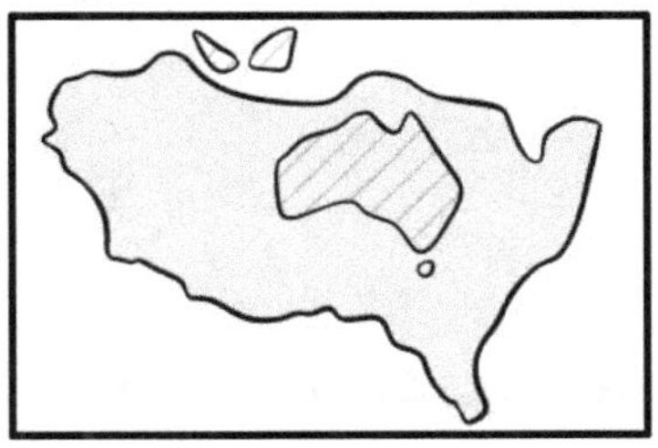

Eurasian Plate

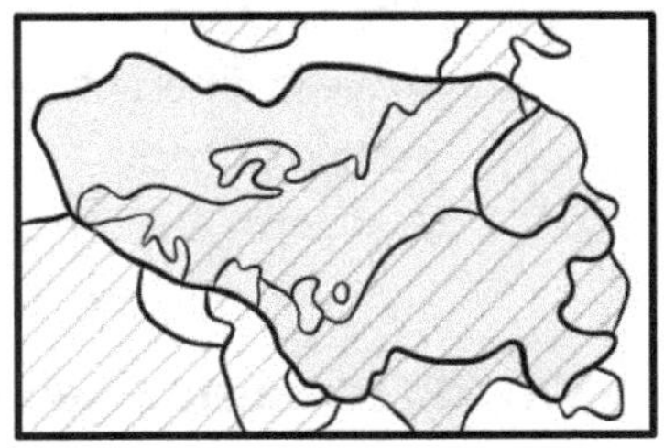

South American Plate

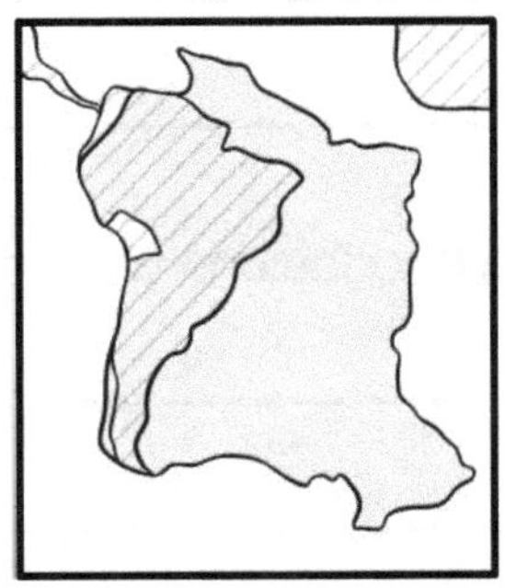

PANGEA

Over 200 million years ago, all the continents we know today were joined together in one huge landmass called **Pangea**. The continents were stuck together like pieces of a giant puzzle. This supercontinent slowly broke apart because of the movement of tectonic plates, a process known as **continental drift**.

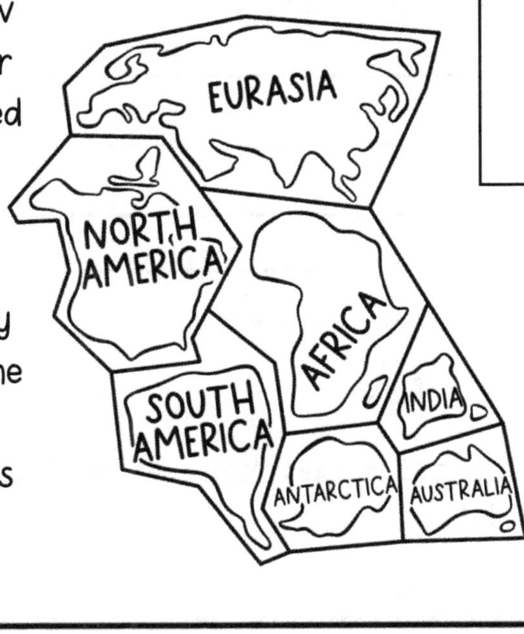

Name: ____________________

Week Eight

Day 1

Plate Tectonics

1. **What are tectonic plates?**

   a) Layers of the atmosphere
   b) Large bodies of water
   c) Pieces of the Earth's crust that fit together like a puzzle
   d) A type of land

2. **How fast do tectonic plates move each year on average?**

   a) 1 to 2 feet
   b) 1 to 2 inches
   c) 1 to 2 miles
   d) They do not move

3. **What is a lithosphere?**

   a) The outermost shell of the Earth made up of the crust and upper part of the mantle
   b) The hot fluid that allows the plates to move in different directions
   c) A type of rock
   d) A layer of the atmosphere

4. **Which type of plate boundary involves two plates moving away from each other?**

   a) Transform
   b) Divergent
   c) Convergent
   d) None of the above

5. **Why do scientists study the movements of tectonic plates?**

____

____

____

6. **Describe one way the movement of tectonic plates can affect people.**

____

____

____

____

7. **True or False: Tectonic plates move very quickly.**

   a) True b) False

8. **True or False: The movements of tectonic plates cause earthquakes, volcanic eruptions, and the formations of mountains.**

   a) True b) False

Week Eight

Day 2

Name: ____________________________

Plate Boundaries

Write a short description for each type of plate boundary that will help you remember what they are and how they work. Then, draw a picture for each type of plate boundary to show how the plates move. If you have internet access, look up the examples listed for each boundary to see how they look!

	DESCRIPTION	ILLUSTRATION	EXAMPLES
TRANSFORM			• San Andreas Fault • New Zealand's Alpine Fault • North America's Queen Charlotte Fault
DIVERGENT			• Mid Atlantic Ridge • Red Sea Rift • Great Rift Valley • East African Rift • West Antarctic Rift
CONVERGENT			• Collision between Eurasian and Indian Plate forming the Himalayas • Subduction of Nazca plate beneath South American Plate to form the Andes

Week Eight

Day 3

Name: ______________________________

Identifying Tectonic Plates

Identify each tectonic plate and write its name next to the corresponding number on the lines provided.

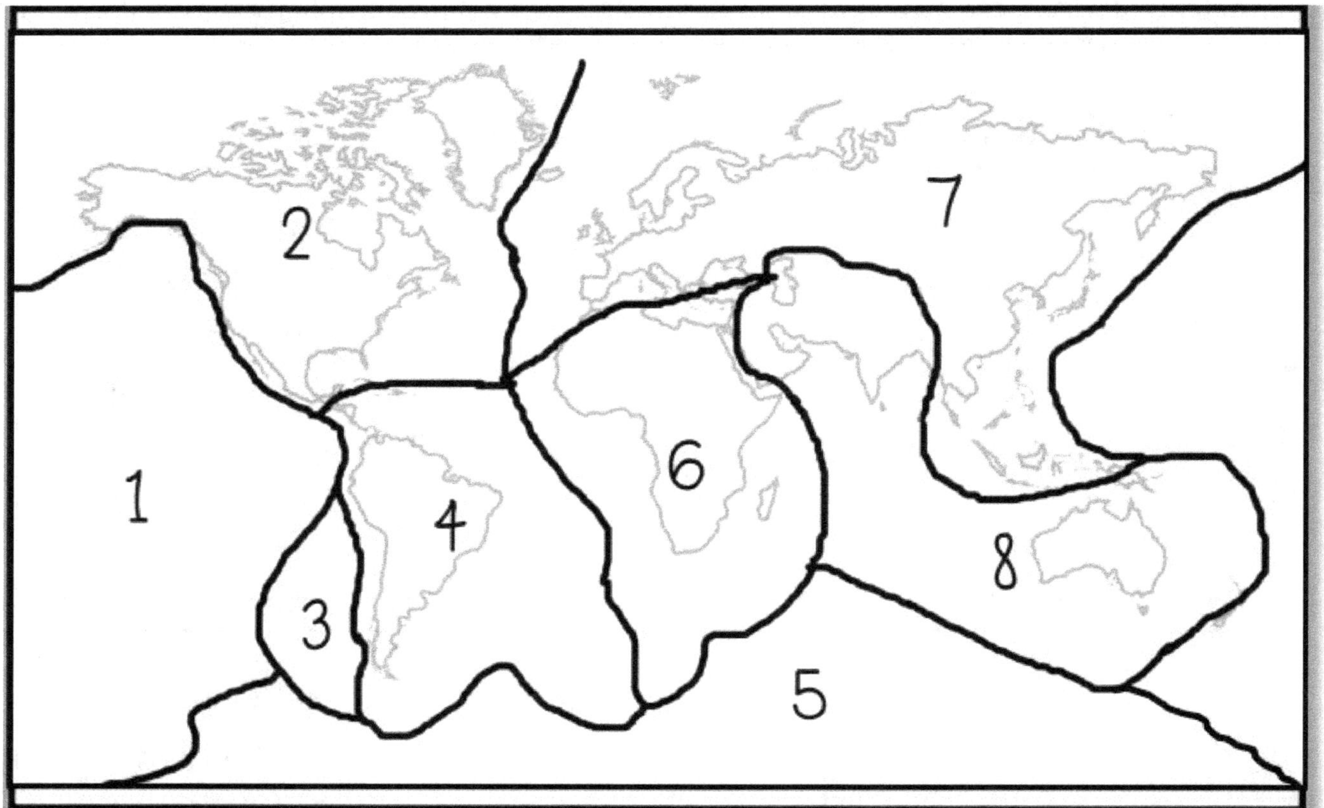

Word Bank

African, Nazca, Australian, North American, Eurasian, Antarctic, Pacific, South American

1.	**5.**
2.	**6.**
3. Nazca Plate	**7.**
4.	**8.**

Name: ____________________

Week Eight

Day 4

Vocab Match

Match each word to its definition. Write the correct letter in the space provided.

1. Tectonic Plates ________
2. Continental Drift ________
3. Pangaea ________
4. Asthenosphere ________
5. Lithosphere ________
6. Transform Boundaries ________
7. Divergent Boundaries________
8. Convergent Boundaries ________
9. Subduction ________

A. The single large supercontinent that existed millions of years ago when all of Earth's continents were joined together

B. A place where two tectonic plates move away from each other

C. When one tectonic plate sinks beneath another plate

D. A place where two tectonic plates slide past each other

E. Large pieces of the Earth's lithosphere that move slowly on top of the asthenosphere

F. The layer of the Earth's mantle located just below the lithosphere

G. The outer layer of the Earth made up of the crust and the uppermost part of the mantle

H. The slow movement of Earth's continents across the planet's surface over long periods of time

I. A place where two tectonic plates move toward each other and collide

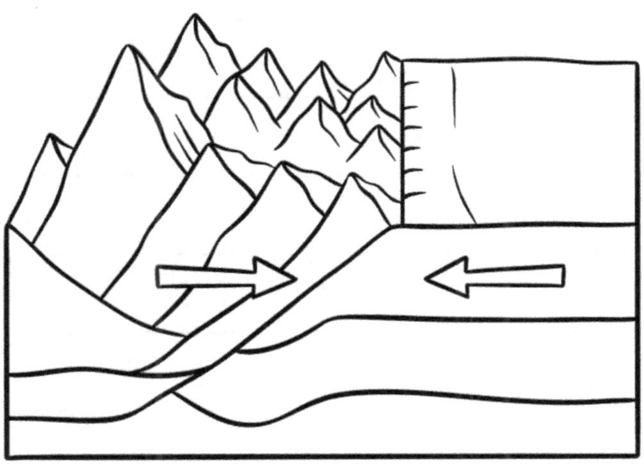

Week Eight

Day 5

1. If tectonic plates move so slowly, why do they still cause such big events like earthquakes and volcanoes?

2. Imagine what would happen if tectonic plates stopped moving. How do you think that would affect the Earth's surface and living things?

Week Nine

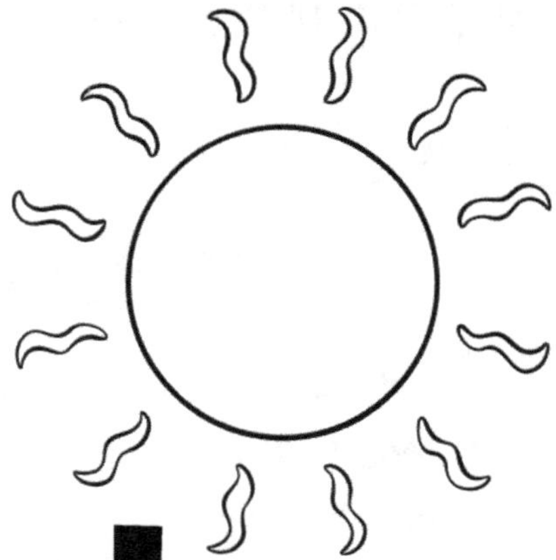

Natural Resources

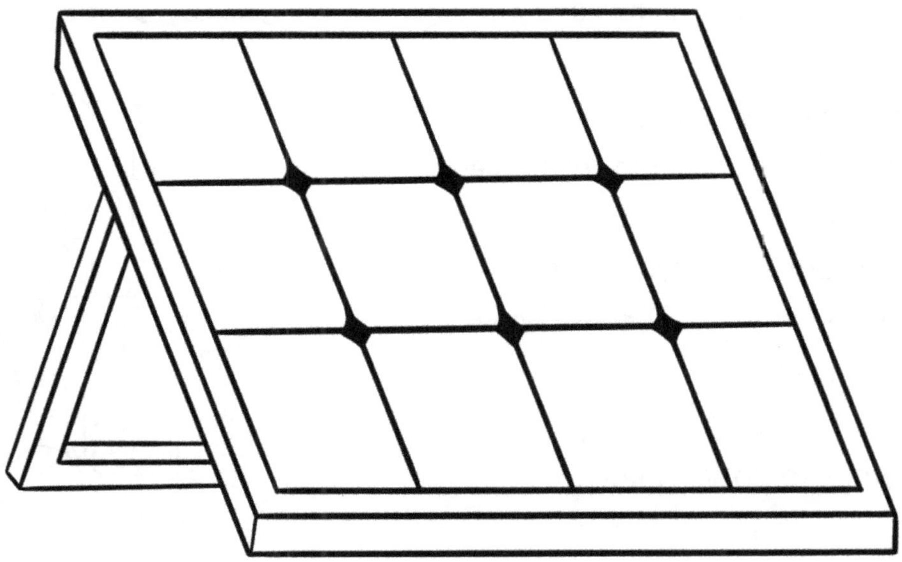

The sun provides enough energy in one hour to power the entire world for a year! Solar panels capture this energy and convert it into electricity.

Name: ____________________

Week Nine

Understanding Natural Resources

Natural resources are the basis of life on Earth. Humans have always depended on things that exist freely in nature to survive. People do not make natural resources. Instead, they gather them from the Earth. Examples include sunlight, air, water, wood, oil, wind power, hydroelectric energy, and coal.

There are two types of natural resources: renewable resources and nonrenewable resources. **Renewable resources** are resources that can be replaced or renewed naturally and will not run out. Examples include soil, sunlight, water, and wood. When a tree is cut down, another tree can grow in its place, making trees a renewable resource.

Nonrenewable resources, on the other hand, will eventually run out and take a very long time to be replaced. Examples of nonrenewable resources include fossil fuels like coal, oil, and natural gas. Did you know that the gasoline we use in our cars will eventually run out? Over time, people will need to find other ways to power cars.

People cannot live without natural resources because they depend on them every day. For example, you need resources to eat and drink. The silverware you use to eat with is made from raw materials that come from natural resources. Think about all the buildings in your neighborhood that need energy for heating and cooling. Natural resources are everywhere, and how we use them today will help shape our future.

Name: ______________________________

Biomass Energy

Biomass is material made from living things such as plants and animals that we can convert into energy.

Wood

Crops

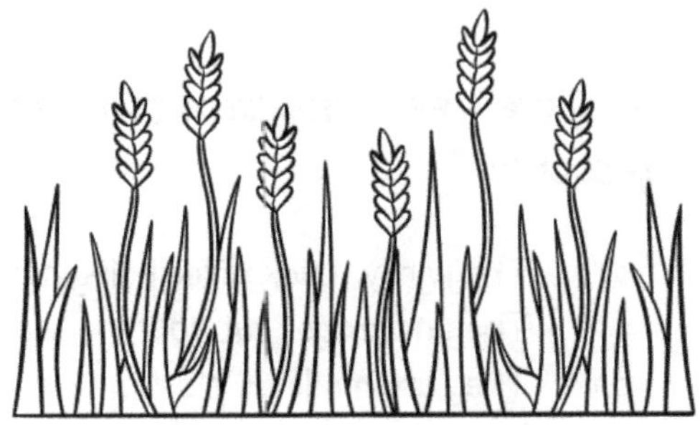

Types of Biomass

Garbage

Animal Waste

Biomass is considered a renewable energy source because trees and crops can be grown again, and organic waste can be reused to produce energy.

Name: ______________________

Week Nine

Day 1

Natural Resources

1. **What are natural resources?**

   a) Things people create from raw materials
   b) Things that exist freely in nature and are gathered from the Earth
   c) Man-made items used for building
   d) Artificial sources of energy

2. **Which of the following is a renewable resource?**

   a) Coal
   b) Oil
   c) Natural gas
   d) Sunlight

3. **Why are trees considered renewable resources?**

   a) They grow back after being cut down
   b) They never get cut down
   c) They are made of plastic
   d) They do not need water to grow

4. **What will happen if we continue to use non-renewable resources?**

   a) They will never run out
   b) They will eventually run out
   c) They will replace themselves quickly
   d) They will turn into renewable resources

5. **Explain the difference between renewable and non-renewable resources in your own words.**

____

____

____

6. **Why is it important to manage how we use natural resources?**

____

____

____

____

7. **People make natural resources and gather them from the Earth.**

   a) True   b) False

8. **The energy used to heat and cool buildings come from natural resources.**

   a) True   b) False

Name: ______________________

Coal Formation

Coal is a type of fossil fuel that people use to generate electricity. It is a black rock made mostly from the remains of dead plants that lived millions of years ago. Over a very long time, these plant remains changed into coal. Did you know that plants can eventually turn into rock?

Millions of years ago, large plants grew in warm, swampy areas. When these plants died, their remains sank to the bottom of the swamp. Over time, layers of water, dirt, and rock covered the dead plants. New plants continued to grow on top, and when they died, they added even more layers.

As the layers became thicker, heat and pressure slowly built up. Over millions of years, this heat and pressure changed the dead plant material into coal.

Coal formation is a process that still occurs today, but it takes millions of years to complete.

1. Illustrate and describe the coal formation process in 3 steps:

STEP 1	STEP 2	STEP 3

2. Is coal a renewable or a non-renewable resource? Explain why.

____

____

Week Nine

Day 3

Name: ______________________

Renewable or Non-Renewable

Look at each picture. Decide whether it shows a renewable or non-renewable resource.
Write renewable or non-renewable on the line below each picture.

1. WIND POWER

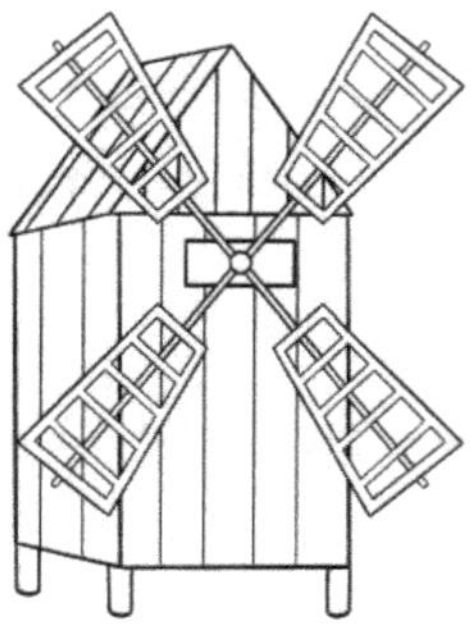

____

2. BIOMASS

____

3. NATURAL GAS

____

4. SOLAR ENERGY

____

5. OIL

____

6. ROCKS & MINERALS

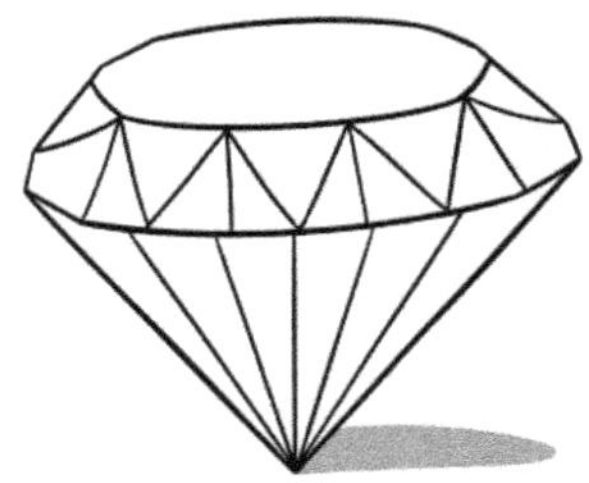

____

7. GEOTHERMAL ENERGY

____

8. HYDROELECTRIC POWER

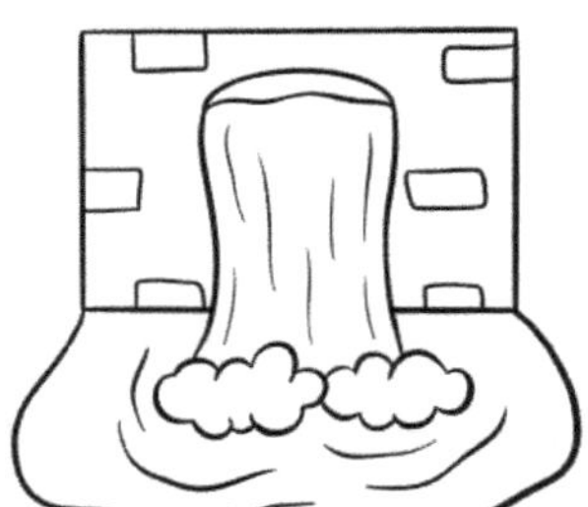

____

9. COAL

____

Name: ______________________

Week Nine

Day 4

Vocab Match

Match each word to its definition. Write the correct letter in the space provided.

1. Natural Resource ________
2. Renewable Resource ________
3. Non-renewable Resource ________
4. Biomass ________
5. Conservation ________
6. Fossil Fuels ________
7. Hydroelectric Energy ________
8. Geothermal Energy ________

A. Electricity produced by moving water

B. An energy resource that can be replaced naturally and does not run out

C. The careful use of natural resources so that they can last as long as possible

D. Energy produced by heat from inside the Earth

E. An energy resource that cannot be replaced as quickly as it is used and takes millions of years to form

F. Includes oil, coal, and natural gas formed from the remains of plants and animals millions of years ago

G. Materials we get from nature, such as air and water, that people use to meet their needs

H. Any material made by plants and animals that we can convert into energy

Week Nine

Day 5

How can human activities impact the availability of natural resources, and what can we do to protect these resources?

Exploring Space Science

Week One

The PLANETS

Jupiter is so big that you could fit over 1,300 Earths inside it. This makes it the largest planet in our solar system!

Week One

Read the passage to complete the tasks for the next few days.

EXPLORING THE PLANETS

We have 8 planets in our solar system that orbit, or circle, the Sun.

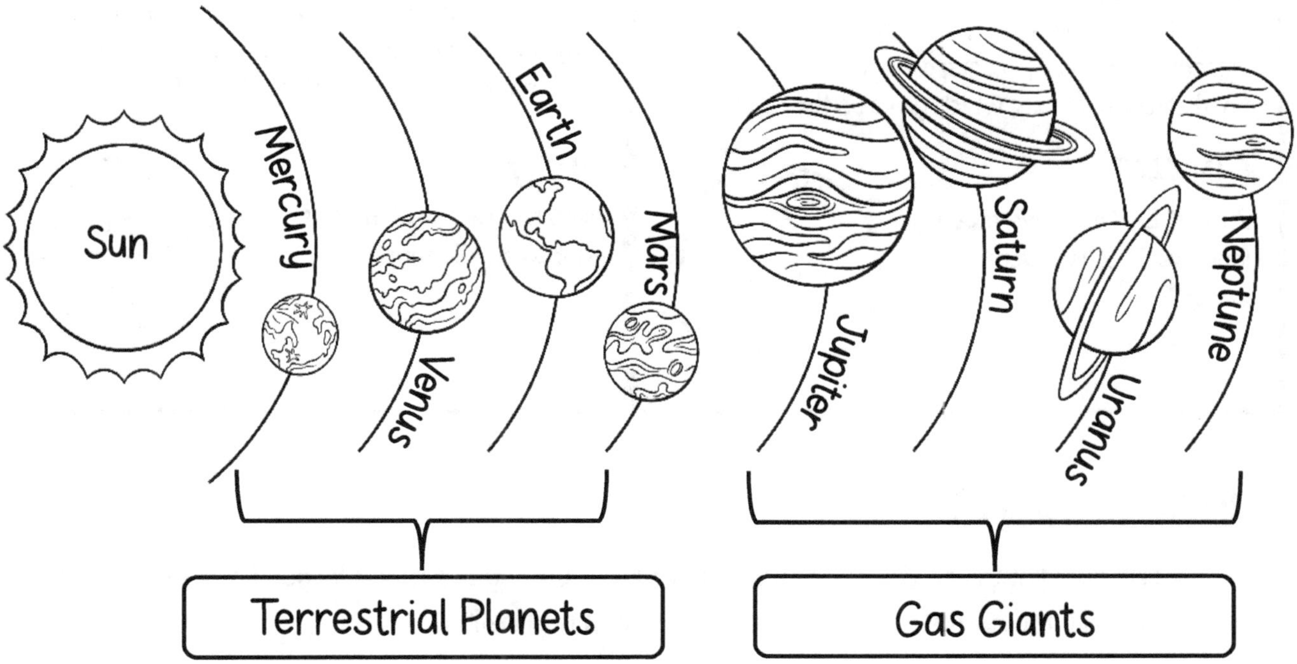

THE TERRESTRIAL PLANETS

The terrestrial planets, also known as the rocky planets, are the four innermost planets of the solar system. They are made of rock and metal and have solid surfaces.

MERCURY

Mercury is the smallest planet and closest to the Sun. It has a very thin atmosphere, which means it cannot retain heat. So, its surface experiences extreme temperatures, both hot and cold.

FUN FACT

A year on Mercury (one orbit around the Sun) takes just 88 Earth days!

VENUS

Venus is known as Earth's sister because it is almost identical in size, composition, gravity, and density. Unlike Earth, the atmosphere on Venus is mostly carbon dioxide. It also has clouds of sulfuric acid that trap heat, making it the hottest planet in the solar system. Venus also has volcanoes, mountains, and vast plateaus.

FUN FACT

Venus rotates very slowly and in the opposite direction to most planets, meaning a day on Venus is longer than a year!

EARTH

Earth is the third planet from the Sun and the only one known to support life. It is the only planet that has 21% oxygen which make life possible. It has a perfect mix of air, water, and land, making it a vibrant place full of plants, animals, and humans. Earth tilts on its axis, which gives us the seasons.

FUN FACT

29% of Earth is covered with land and 71 % is covered by water.

MARS

Mars is known as the Red Planet due to its reddish appearance, which comes from iron oxide (rust) on its surface. It has the tallest volcano and the deepest canyon in the solar system. Mars lacks oxygen for humans, has many dust storms, and does not have liquid water on its surface.

FUN FACT

Mars has two small moons, Phobos and Deimos.

THE GAS GIANTS

The gas giants are the four outermost planets. They are much larger than terrestrial planets and are made mostly of gases like hydrogen and helium.

JUPITER

Jupiter is the largest planet in our solar system. It has a Great Red Spot, which is a massive storm bigger than Earth. Jupiter spins around much faster than Earth, so a day and night cycle lasts about 9.84 hours rather than 24 hours! In fact, Jupiter is the fastest spinning planet in the solar system. Thick colorful clouds surrounding Jupiter are made mostly of poisonous gases like ammonia, crystals, and sulfur.

FUN FACT

Jupiter has at least 79 moons.

SATURN

Saturn is the second-largest planet in our solar system. It is a great ball of hydrogen and helium. Saturn is famous for its beautiful ring system, made up of ice and rock particles. It has at least 18 moons that orbit around the planet.

FUN FACT

Saturn's largest moon, Titan, is bigger than the planet Mercury and has a thick atmosphere.

Week One

URANUS

Uranus is the 7th planet from the Sun. It was the first planet discovered by a telescope. Uranus is unique because it rotates on its side, making its axis nearly horizontal. It has a faint ring system and is known for its blue-green color due to methane in its atmosphere.

FUN FACT

Uranus has 27 known moons, named after characters from the works of Shakespeare and Alexander Pope.

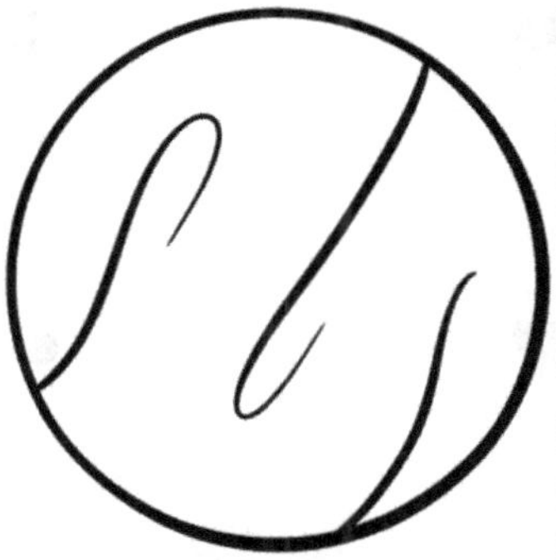

NEPTUNE

Neptune is the farthest planet from the Sun and is known for its deep blue color. It has strong winds and storms, including the Great Dark Spot, similar to Jupiter's Great Red Spot. In fact, Neptune has the windiest weather in the solar system. Wind speeds on Neptune can reach up to 1,500 miles per hour, which is faster than the speed of sound on Earth.

FUN FACT

Neptune's moon Triton is one of the coldest places in our solar system and has geysers that spew nitrogen gas.

Week One

Day 1

Name: ______________________

WHICH PLANET AM I?

After reading and learning all about the fascinating planets in our solar system, can you identify which planet is being described? Draw a line from each description to the correct planet.

1 I was the first planet discovered by a telescope and I rotate on my side. Who am I?	JUPITER
2 I am the smallest planet in our solar system and am the closest planet to the Sun. Who am I?	VENUS
3 71% of my surface is covered with water and I have one moon. Who am I?	URANUS
4 I am the second-largest planet and am famous for my beautiful rings made up of ice and rock. Who am I?	NEPTUNE
5 I am the largest and fastest-spinning planet in the solar system. Who am I?	SATURN
6 I am known as the Red Planet and have the tallest volcano in the solar system. Who am I?	EARTH
7 I am the farthest planet in our solar system and have the windiest weather. Who am I?	MERCURY
8 I am the hottest planet and I spin backwards compared to most planets. Who am I?	MARS

THE PLANETS

Week One

Day 2

Name: ______________________________

List the planets in the correct order and indicate whether each one is a terrestrial planet or a gas giant. Write "Terrestrial" or "Gas Giant" on the line next to each planet.

WORD BANK

Mars	Neptune	Jupiter	Mercury	Venus	Saturn	Uranus	Earth

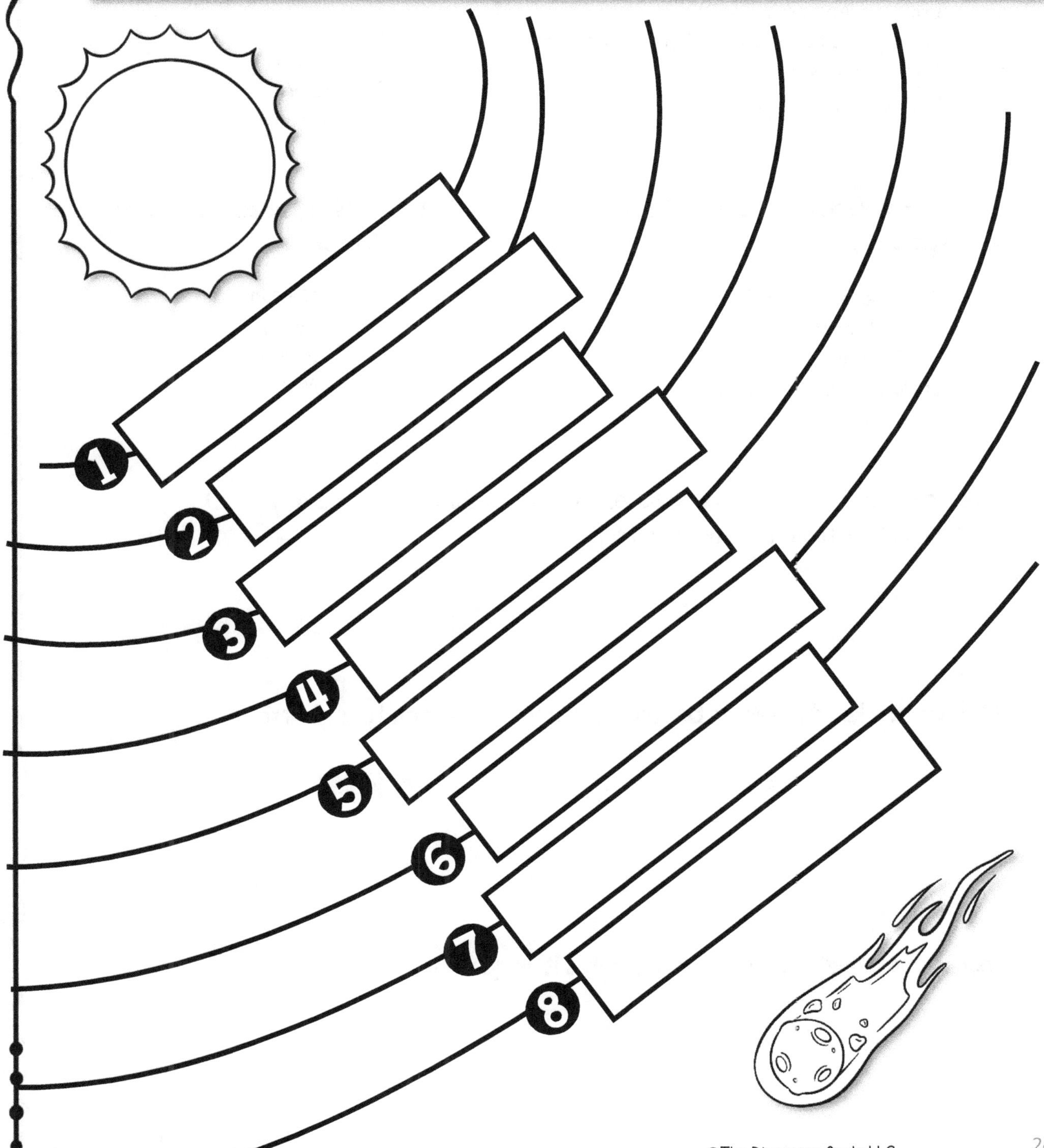

Name: ______________________

EXPLORING THE PLANETS

After learning about the planets, answer the following questions.

1. **What is unique about Uranus's rotation?**

   a) It rotates very quickly
   b) It rotates the opposite direction of most planets
   c) It rotates on its side
   d) It rotates very slowly

2. **Which planet is known for having the Great Red Spot?**

   a) Mars
   b) Jupiter
   c) Saturn
   d) Neptune

3. **What makes Venus the hottest planet in our solar system?**

   a) Its closeness to the Sun
   b) Its thick, carbon dioxide–rich atmosphere
   c) Its fast rotation
   d) Its large number of volcanoes

4. **Which planet has a moon named Triton?**

   a) Neptune
   b) Saturn
   c) Jupiter
   d) Mars

5. **True or False: (If your answer is false, rewrite the statement to make it true.) Earth is farther from the Sun than Mars.**

6. **What are the key differences between terrestrial planets and gas giants?**

7. **Choose one planet and explain what makes it unique.**

Name: ______________________

Week One

Day 4

Vocab Match

Match each word with its definition. Write the correct letter in the space provided.

1. Terrestrial Planets ________
2. Gas Giants ________
3. Orbit ________
4. Atmosphere ________
5. Gravity ________
6. Rotation ________
7. Revolution ________
8. Axis ________
9. Moons ________
10. Solar System ________

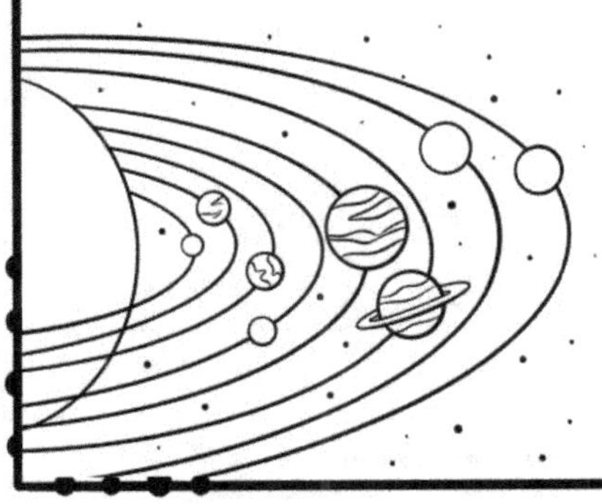

A — An imaginary line that a planet spins around, causing day and night

B — The path a planet follows as it moves around the Sun

C — Natural satellites that orbit planets

D — Large planets made mostly of gases, including Jupiter, Saturn, Uranus, and Neptune

E — The force that pulls objects toward the center of a planet

F — Rocky planets that are close to the Sun, such as Mercury, Venus, Earth, and Mars

G — The movement of a planet as it travels around the Sun

H — The spinning of a planet on its axis

I — The Sun and all the objects that orbit it, including planets, moons, asteroids, and comets

J — The layer of gases surrounding a planet

Week One

Day 5

If you were planning a mission to another planet, which planet would you choose and why? Use facts from what you learned this week to support your answer.

Week Two

Moon PHASES

The Moon's phases change as it orbits Earth, from a new moon that's completely dark to a full moon that lights up the night sky like a giant, glowing lantern!

PHASES OF THE MOON

Read the passage to complete the tasks for the next few days.

As the Moon orbits Earth, the Sun lights up different parts of the Moon, making it seem as if the Moon is changing shape. The Moon's shape never changes. We can only see the Sun's light reflected off the Moon's surface. The different shapes of the Moon that we see from Earth are called the **phases of the Moon.**

It takes the Moon a lunar month, or about 29.5 days, to go through its eight different phases. When the bright part of the Moon is getting bigger, the Moon is **waxing**. When the bright part is getting smaller, the Moon is **waning**.

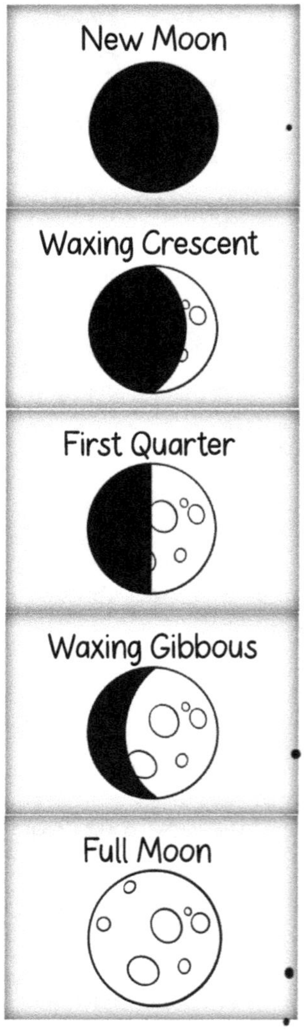

During the **New Moon** phase, the Moon is between the Earth and the Sun. We cannot see a New Moon because the side of the Moon facing Earth is not lit by the Sun.

A few days after the New Moon, a **Waxing Crescent** can be seen as a thin silver of light. It increases, or "waxes", in size from one day to the next.

When the Moon's right half is lit by the sun, it is called the **First Quarter**. Sometimes, this phase is called a Half Moon. You should be able to see a First Quarter Moon about one week after a New Moon.

During the following week, as the Moon continues to wax, it becomes more than half lit. This phase is called **Waxing Gibbous**.

When the Moon has completed half of its trip around Earth from the previous New Moon, a **Full Moon** occurs. During a Full Moon, we can see the entire lit portion of the Moon because the side facing Earth is completely turned toward the Sun.

Phases of the Moon

After the Full Moon, the Moon begins to **wane**, or decrease in size. The Moon is now in the **Waning Gibbous** phase. Darkness starts to creep in from the right as the left side remains lit..

Waning Gibbous

Next, a **Last Quarter** Moon occurs when the left half of the Moon is lit.

Last Quarter

Finally, during the last week of its monthly orbit, a **Waning Crescent** appears as the Moon's left side becomes a thin silver of light.

Waning Crescent

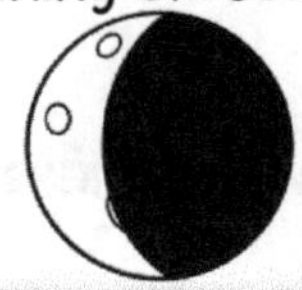

The Moon will become a **New Moon** again a day or two after a Waning Crescent, and the cycle starts all over again!

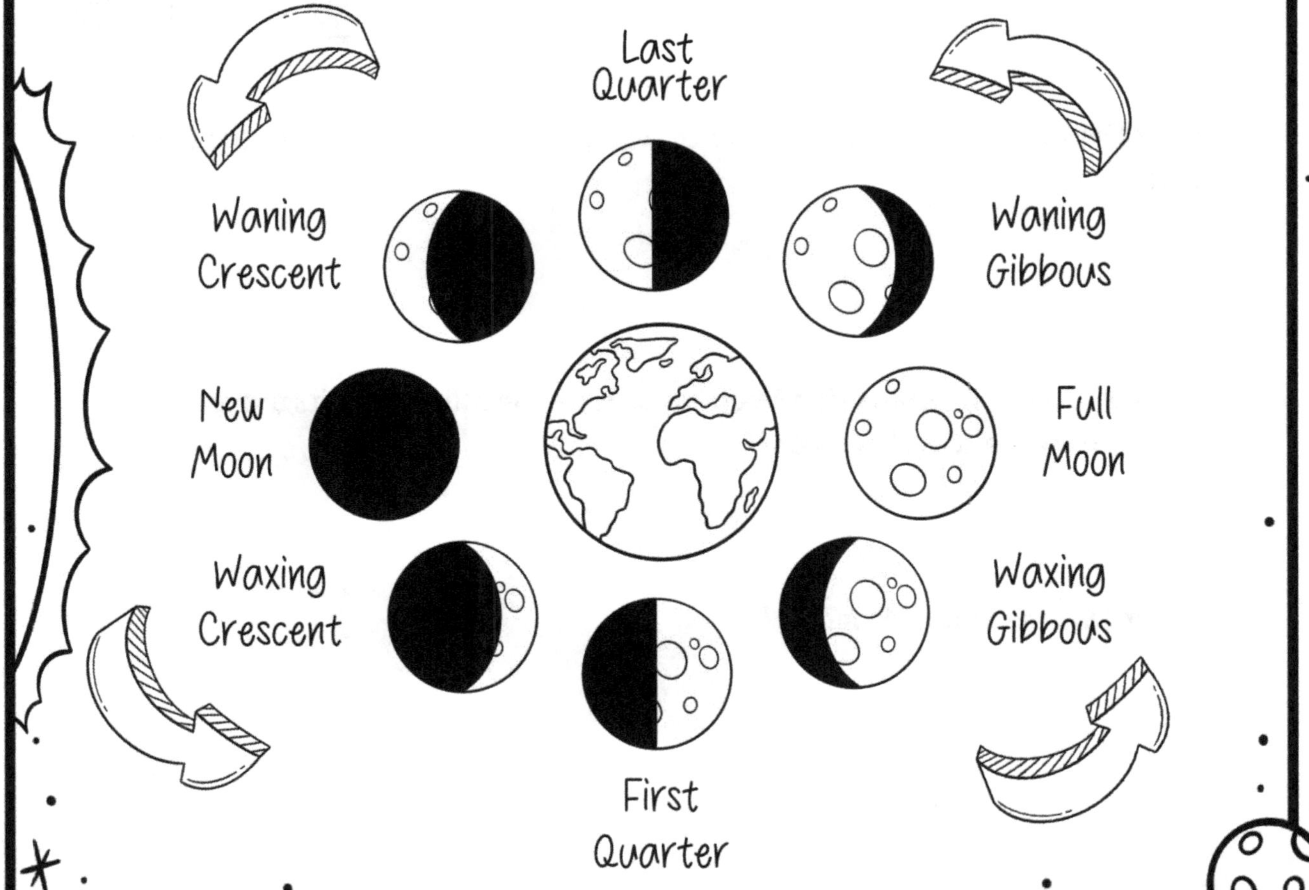

Name: ______________________

PHASES OF THE MOON

After learning about the phases of the Moon, answer the following questions.

1. **Why does the Moon's shape appear to be changing?**
   a) Because it is changing shape
   b) Due to the Moon's orbit around the Earth
   c) Because of the Earth's shadow
   d) Due to the moon's brightness

2. **How long does it take the Moon to go through the eight different phases?**
   a) 7 days
   b) 14 days
   c) 29.5 days
   d) 365 days

3. **Which phase comes after a waxing crescent?**
   a) New Moon
   b) First Quarter
   c) Waxing Gibbous
   d) Full Moon

4. **Which picture shows a Waning Gibbous Moon?**

a) 

b) 

c) 

d) 

5. **True or False: A waxing Moon appears to be getting smaller. (If false, rewrite the sentence to make it true.)**

6. **What is the Moon phase when the Moon has completed exactly half of its trip around the Earth from the previous New Moon?**

7. **Why can't we see a New Moon?**

Name: ______________________________

Week Two

Day 2

MOON PHASES DIAGRAM

Label each phase of the Moon using the word bank.

Waning Gibbous	New Moon	Full Moon	Last Quarter
Waxing Crescent	Waxing Gibbous	Waning Crescent	First Quarter

Week Two

Day 3

Name: ______________________________

LET'S SHADE!

Shade in the circle to show the Moon phase that is written.

1. 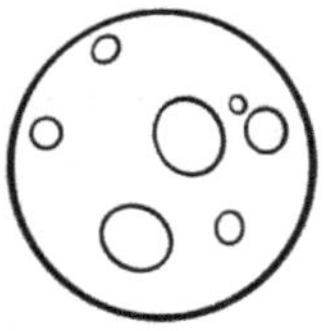
Full Moon

2. 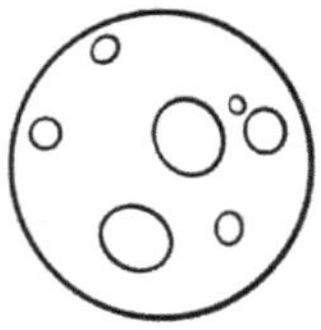
New Moon

3. 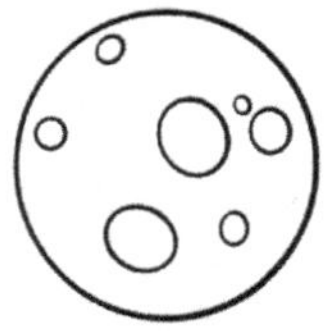
First quarter

4. 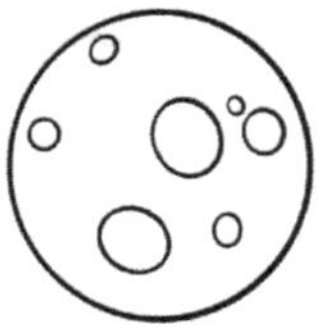
Waxing Gibbous

5. 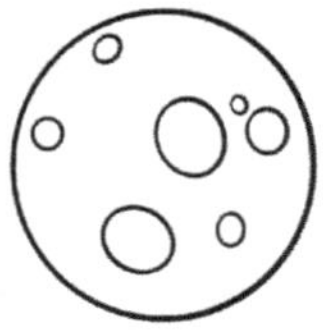
Waning Crescent

6. 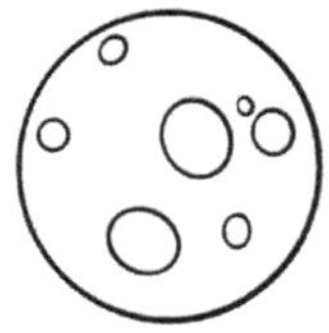
Last Quarter

7. 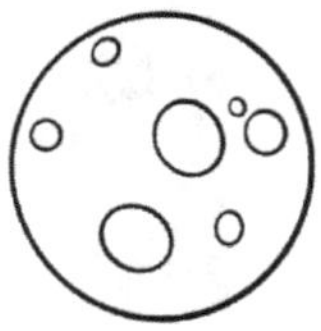
Waxing Crescent

8. 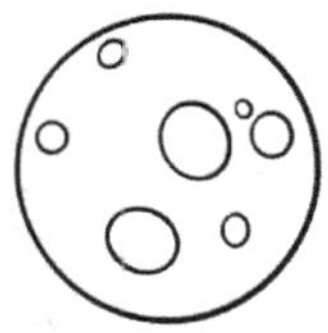
Waning Gibbous

---

9. **Is the Moon waxing or waning?**

a) Waxing

b) Waning

10. **Is the Moon waxing or waning?**

a) Waxing

b) Waning

Name: ____________________

Week Two

Day 4

VOCAB MATCH

Match each word to its definition. Write the correct letter in the space provided.

1. Moon ________
2. Phase ________
3. Crescent ________
4. Gibbous ________
5. Quarter ________
6. New Moon ________
7. Full Moon ________
8. Waxing ________
9. Waning ________

A. More than half of the Moon is lit
____

B. The list part of the Moon is getting bigger
____

C. One of the different shapes of the Moon seen from Earth
____

D. The Moon is not visible from Earth
____

E. The entire Moon is lit
____

F. Half of the Moon is lit
____

G. The lit part of the Moon is getting smaller
____

H. Earth's natural satellite
____

I. Less than half of the Moon is lit

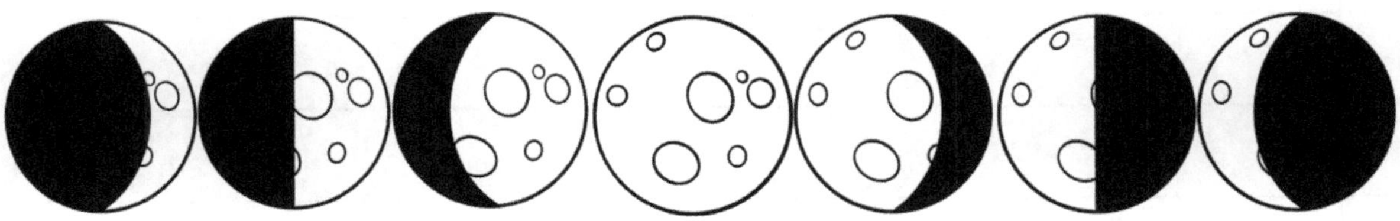

Week Two

Day 5

1. If you observe a full moon tonight, what phase will the moon be in about a week? Why?

2. How would moon phases appear different if you were standing on the Moon looking at Earth?

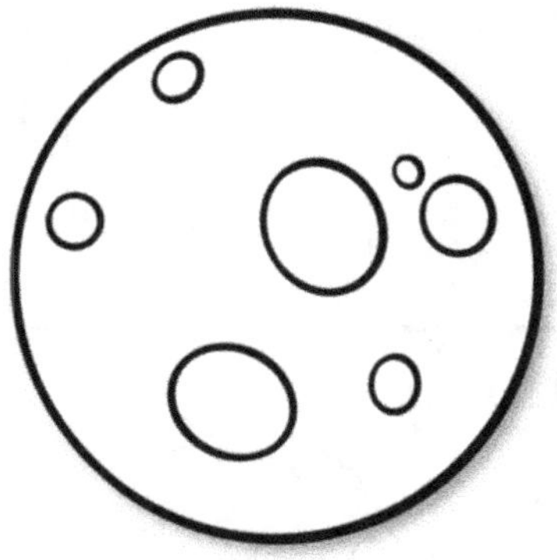

BECOME

a

MOON SPY!

As you watch the Moon night after night, you'll see its phases shift, with each phase revealing a new part of the Moon's surface!

BE A MOON SPY!

Spy on the Moon every night for 30 days. Record the date, shade in the phase you see, and write the phase name.

Date: 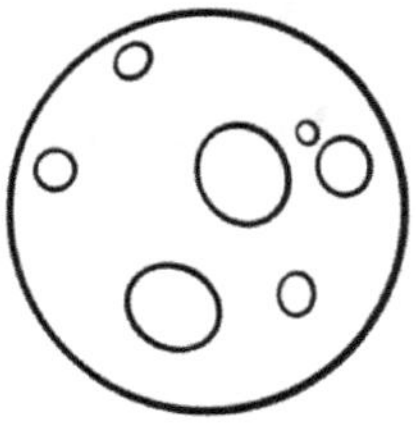 Phase:	Date: 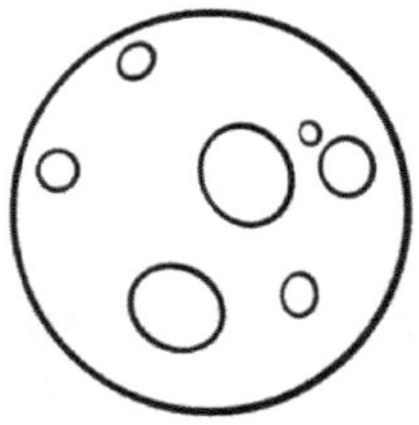 Phase:	Date: 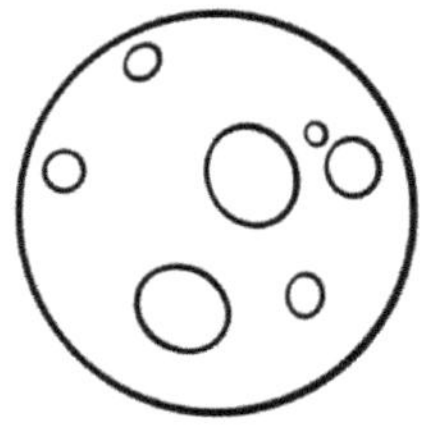 Phase:
Date: 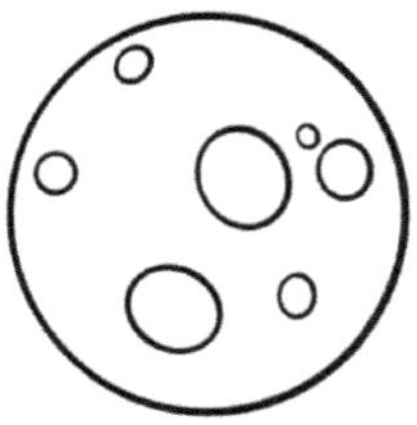 Phase:	Date: 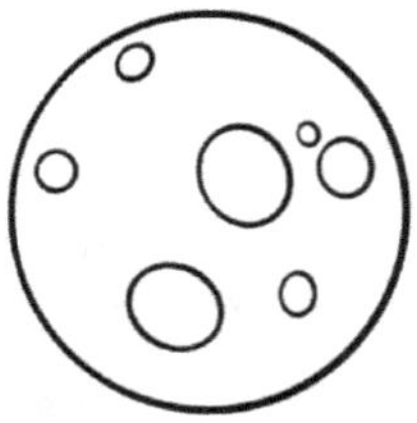 Phase:	Date: 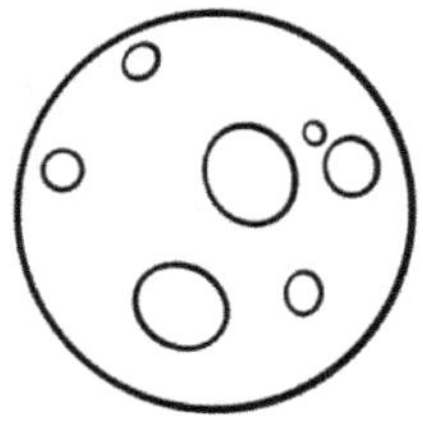 Phase:
Date: 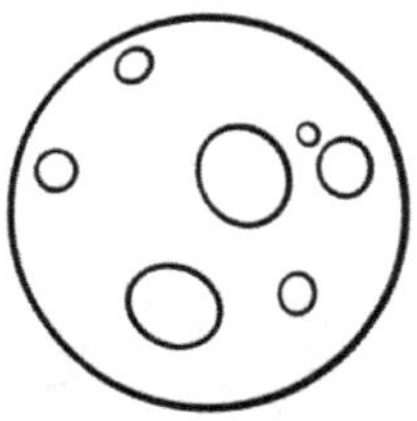 Phase:	Date: 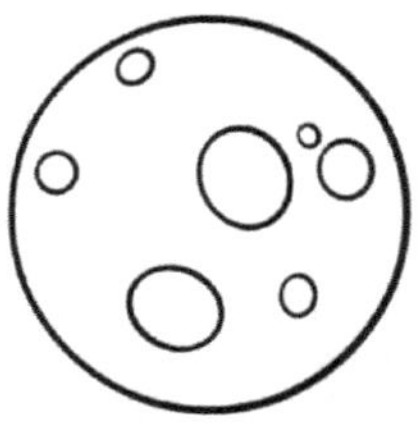 Phase:	Date: 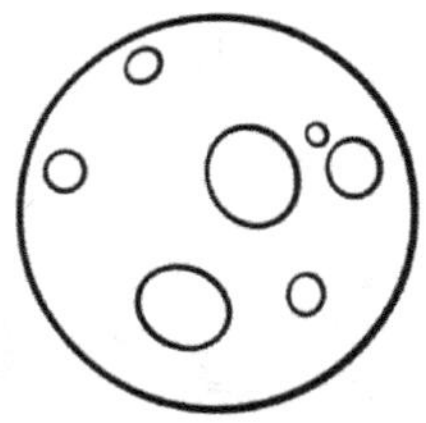 Phase:
Date: 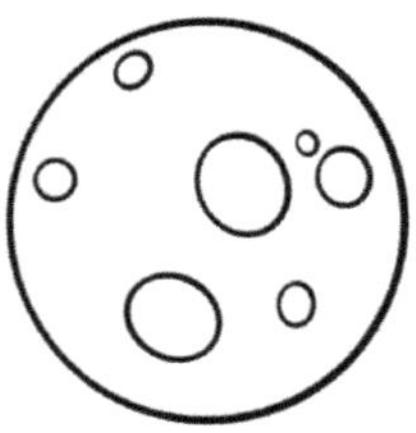 Phase:	Date: 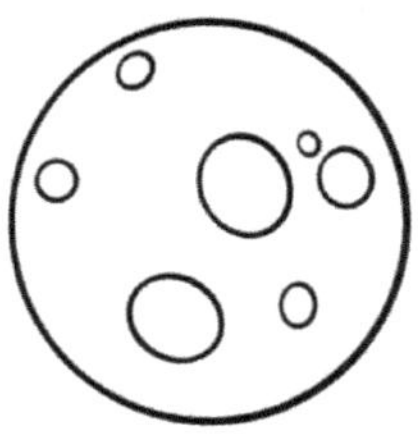 Phase:	Date: 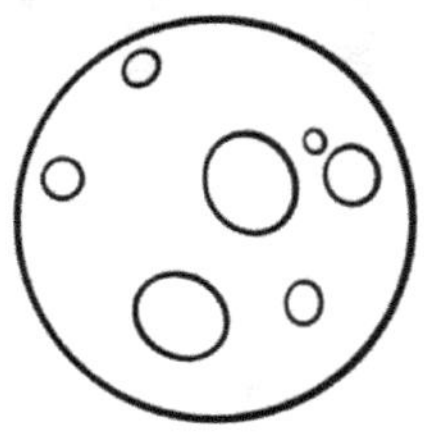 Phase:
Date: 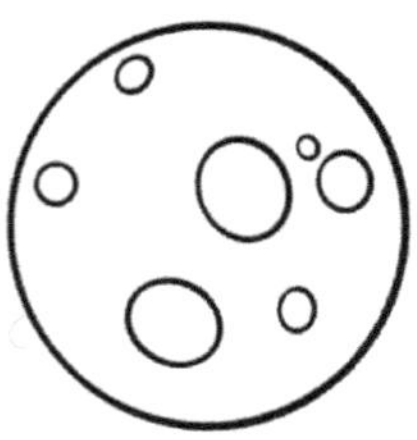 Phase:	Date: 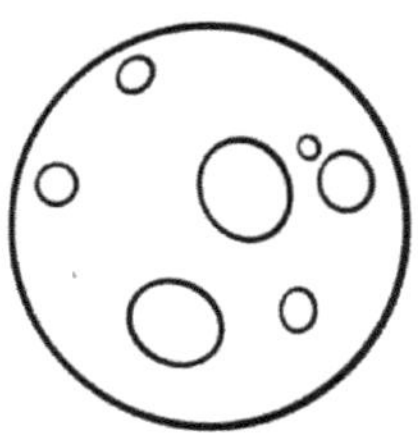 Phase:	Date: 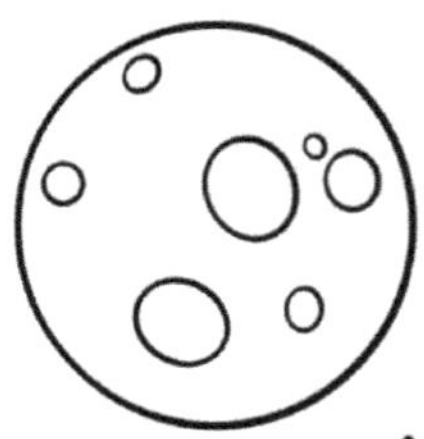 Phase:

Date:

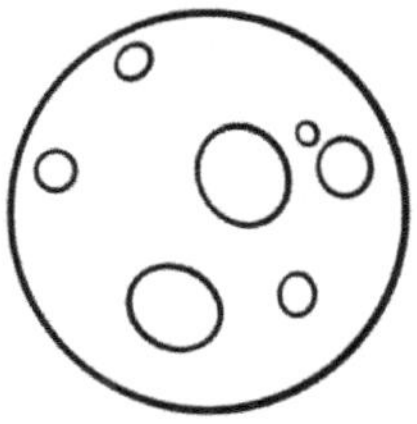

Phase:

Date:

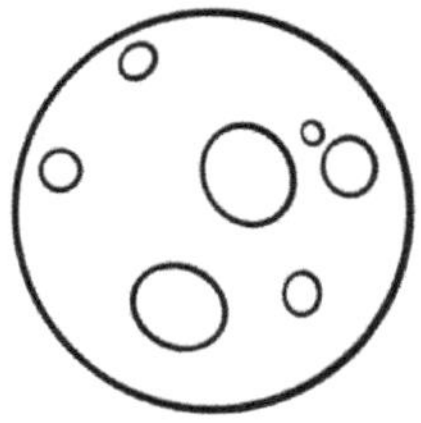

Phase:

Date:

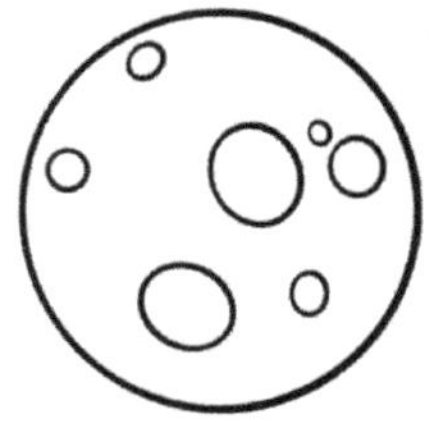

Phase:

Date:

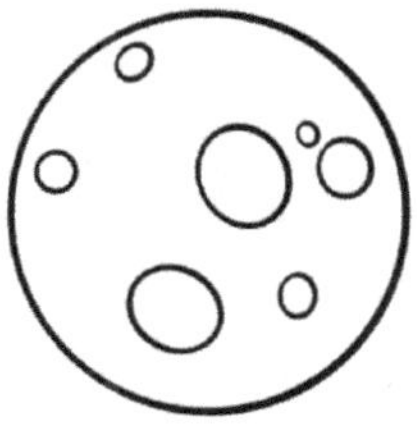

Phase:

Date:

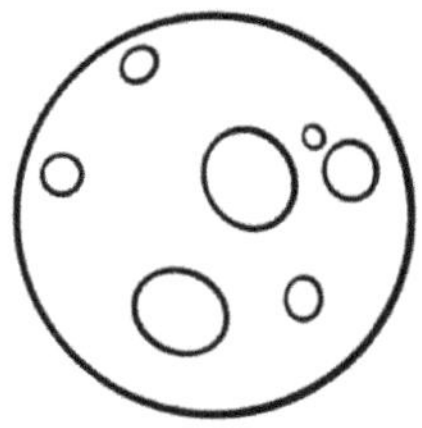

Phase:

Date:

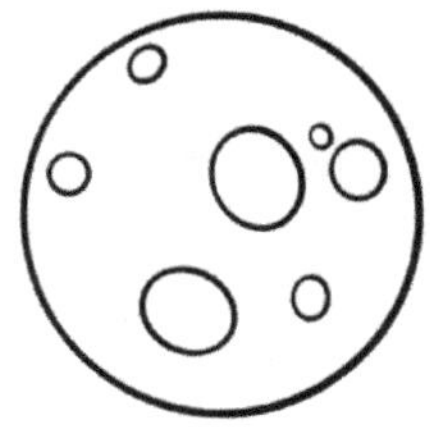

Phase:

Date:

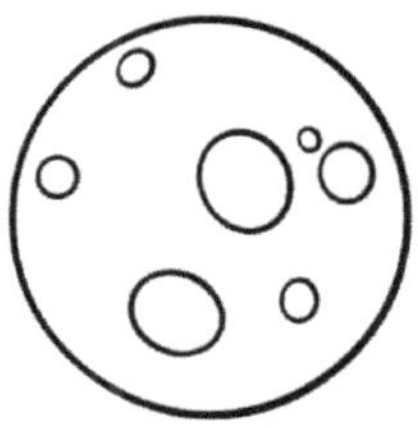

Phase:

Date:

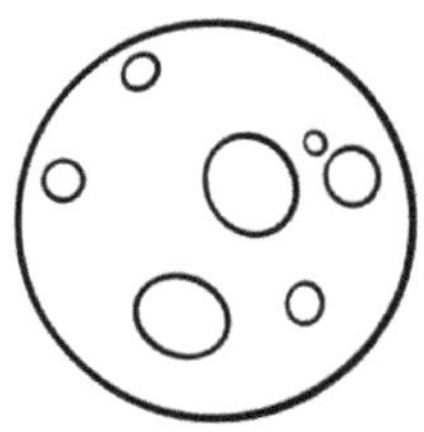

Phase:

Date:

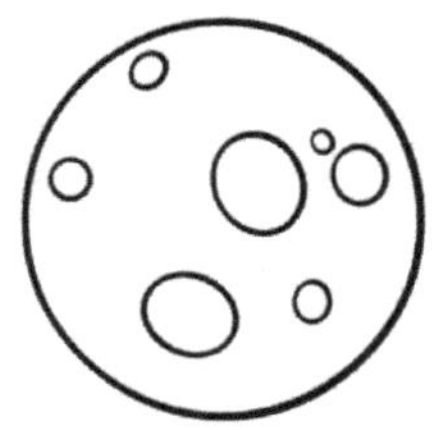

Phase:

Date:

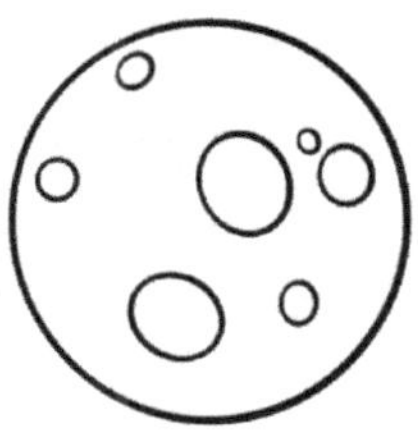

Phase:

Date:

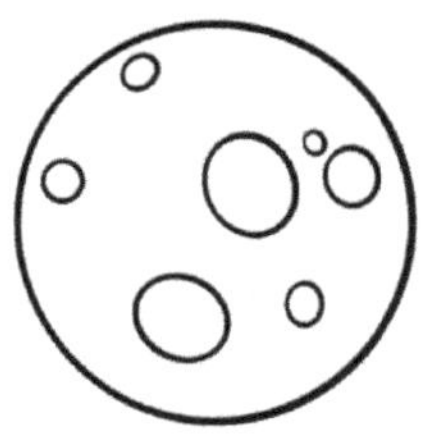

Phase:

Date:

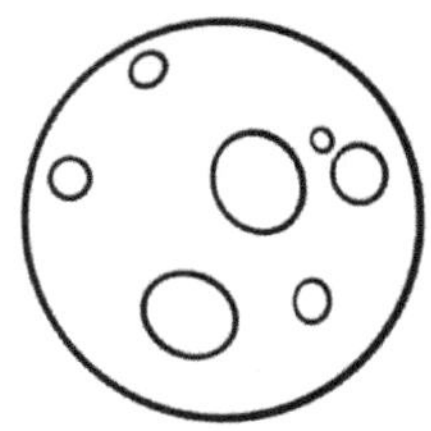

Phase:

Date:

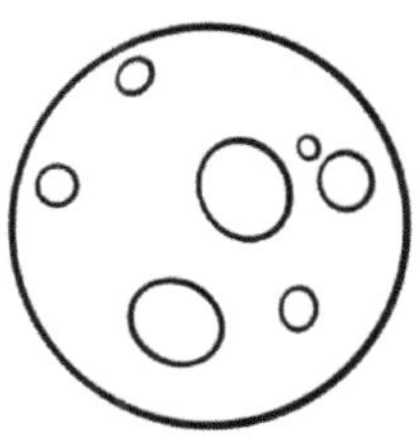

Phase:

Date:

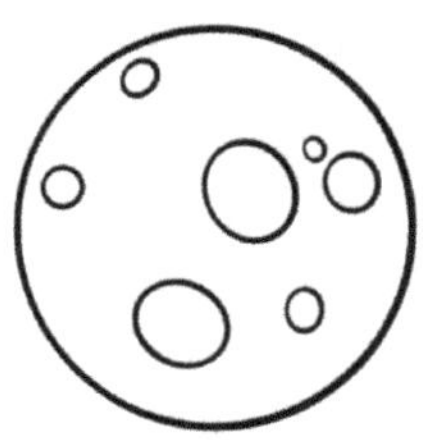

Phase:

Date:

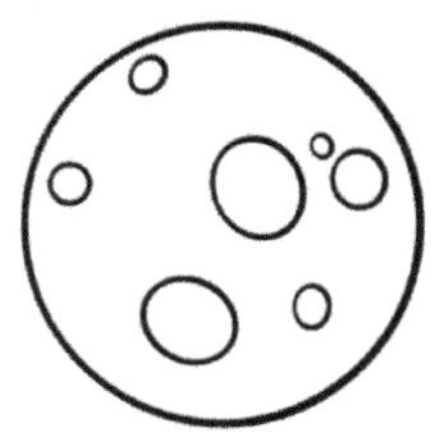

Phase:

Let's Reflect!

Based on your observations, can you predict what the moon will look like on a specific date in the future? Explain.

Week Three

Solar

AND

Lunar

ECLIPSES

During a solar eclipse, the Moon perfectly lines up between the Earth and the Sun, creating a spectacular moment when day briefly turns to night!

Week Three

SOLAR AND LUNAR ECLIPSES

Read the passage to complete the tasks for the next few days.

Have you ever wondered what happens when the Sun, the Moon, and the Earth line up perfectly? When this happens, we get an event called an eclipse. There are two main types of eclipses: **solar eclipses** and **lunar eclipses**. Let's explore both of them.

Solar Eclipse

A solar eclipse happens when the Moon passes between the Earth and the Sun. This blocks the Sun's light and casts a shadow on the Earth. A solar eclipse is not visible from everywhere in the world. It can only be seen from places where the Moon's shadow falls. In these areas, it may appear as though the Sun has gone dark.

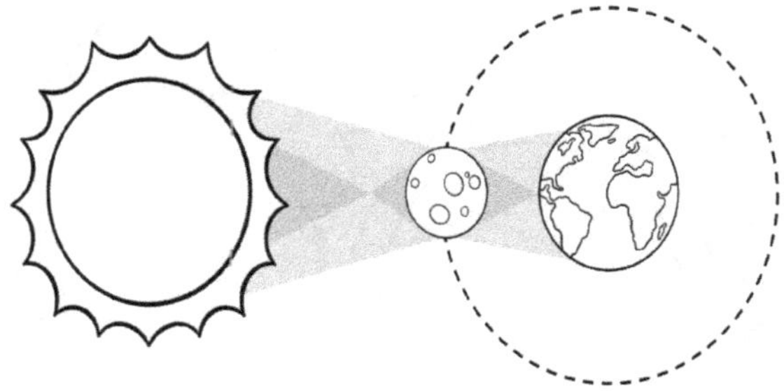

There are three types of solar eclipses:

TOTAL SOLAR ECLIPSE	PARTIAL SOLAR ECLIPSE	ANNULAR SOLAR ECLIPSE
The Moon completely covers the Sun, and it gets dark, like nighttime, for a few minutes. You may see a bright ring of light around the dark Sun, called the **corona**.	Only part of the Sun is covered by the Moon, making it look like a chunk has been taken out of the Sun.	The Moon covers the center of the Sun, leaving a bright ring of light around the edges. This is sometimes called the **"ring of fire."**

Safety Reminder

Never look directly at a solar eclipse without special eclipse glasses. Looking at the Sun can damage your eyes. The Sun's rays are powerful enough to burn the **retina**, the sensitive part at the back of the eye, which can lead to permanent vision loss. Special eclipse glasses are designed to protect your eyes by blocking most of the Sun's harmful rays.

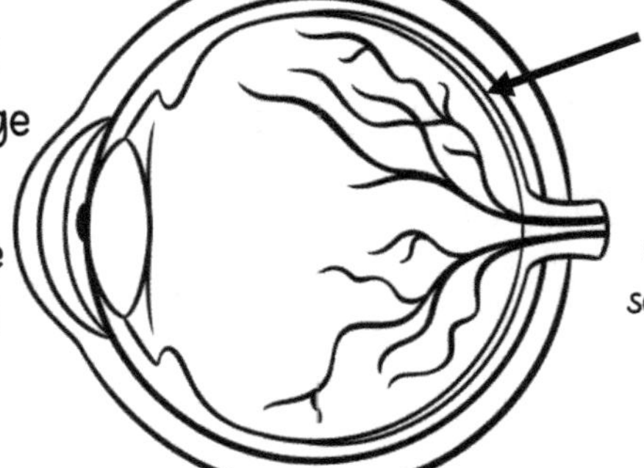

RETINA
The layer of tissue at the back of the eyeball that is sensitive to light

Week Three

Lunar Eclipse

A **lunar eclipse** happens when the Earth moves between the Sun and the Moon. The Earth blocks the Sun's light, and its shadow falls on the Moon. A lunar eclipse can only occur during a **Full Moon.**

Unlike solar eclipses, lunar eclipses can be seen over a much larger area of Earth. Since half of the Earth is always in nighttime, a lunar eclipse can be observed by everyone on that side of the planet.

You can safely view a lunar eclipse without special glasses. During a total lunar eclipse, the Moon often appears reddish in color. This event is sometimes called a "Blood Moon." This happens because some sunlight passes through Earth's atmosphere and is refracted toward the Moon. The light that reaches the Moon is mostly red, which is why the Moon appears dark reddish-brown.

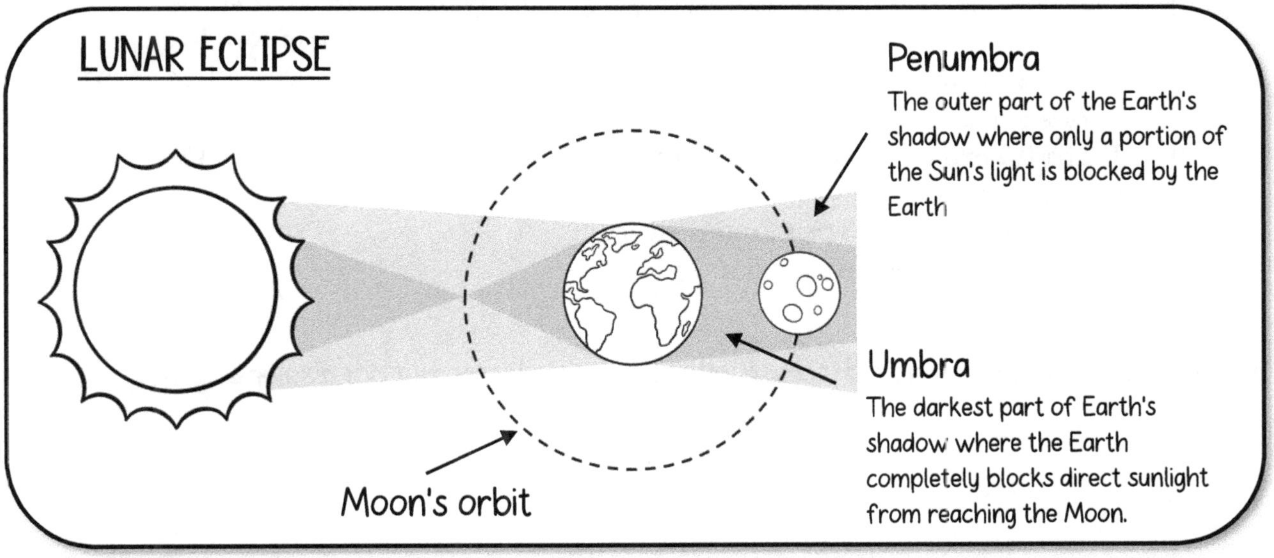

Why Eclipses Are Rare

Eclipses do not happen every month because the orbits of the Earth and the Moon are slightly tilted. The Sun, Earth, and Moon must be perfectly aligned for an eclipse to occur. This alignment only happens a few times each year.

Did You Know?

Long ago, many people were afraid of eclipses because they did not understand what was happening. Some believed eclipses were signs from the gods or warnings that something dangerous or magical was about to occur.

Name: ____________________________

Week Three

Day 1

SOLAR AND LUNAR ECLIPSES

After learning about the solar and lunar eclipses, answer the following questions.

1. **Which of the following is true about a total solar eclipse?**
   a) The Moon turns reddish in color
   b) The Sun's light is completely blocked by the Moon
   c) It can be observed by everyone on the nighttime side of the Earth
   d) It happens only during a Full Moon

2. **What happens during a lunar eclipse?**
   a) The Moon blocks the Earth's shadow
   b) The Moon passes between the Earth and the Sun
   c) The Sun passes between the Earth and the Moon
   d) The Earth passes between the Sun and the Moon

3. **Why do you need special glasses to view a solar eclipse?**
   a) The Sun's rays are too powerful and can damage your eyes
   b) The Moon's rays are too powerful and can damage your eyes
   c) The Earth's rays are too powerful and can damage your eyes
   d) The Sun's and the Moon's rays together are too powerful and can damage your eyes.

4. **What is a "Blood Moon"?**
   a) A total solar eclipse
   b) A type of New Moon
   c) A total lunar eclipse when the Moon appears reddish
   d) A partial lunar eclipse

5. **Why is it safe to view a lunar eclipse without special glasses, but not a solar eclipse?**

6. **Explain why ancient people might have been afraid of eclipses and what they believed these events meant.**

Week Three

Day 2

Name: ______________________________

LABEL IT!

Label the diagram.

WORD BANK

Umbra, Earth, Sun, Penumbra, Moon, Moon's Orbit

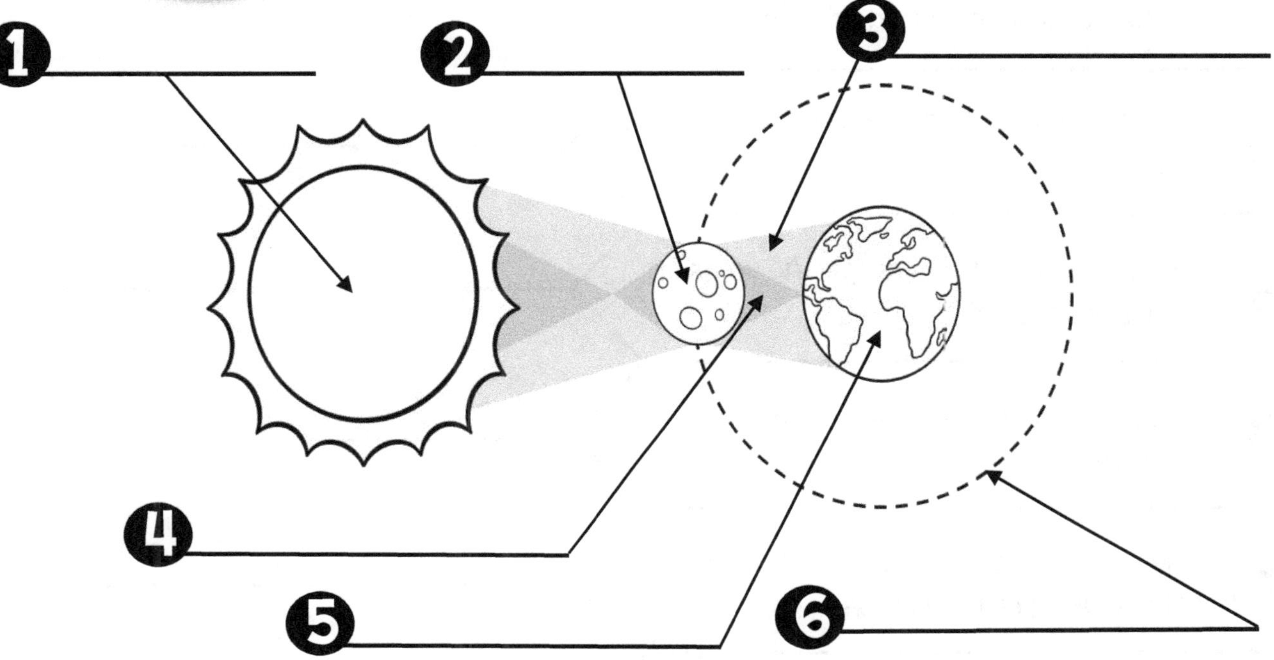

SUMMARIZE IT!

Write a summary of what is happening in the diagram shown above.

7 ______________________________

Week Three

Day 3

Name: ______________________

SOLAR OR LUNAR?

Read each statement carefully. Decide whether it describes a lunar eclipse or a solar eclipse. Write "Lunar" or "Solar" next to each statement.

1 The Sun appears to be partially or completely blocked for a short period of time. ______________________

2 The Earth passes between the Sun and the Moon. ______________________

3. The shadow only covers specific areas on Earth, not the entire planet. ______________________

4. The Moon passes between the Sun and the Earth ______________________

5. The Moon can appear to change color, often looking red or orange. ______________________

6. This type of eclipse can only be seen at night. ______________________

7. This type of eclipse can be viewed without special glasses. ______________________

8. This type of eclipse can only be seen during the day. ______________________

9. This eclipse can only occur during a full moon. ______________________

10. You should not look at this eclipse without special glasses. ______________________

Name: ______________________________

Week Three

Day 4

Vocab Match

Match each word with its definition. Write the correct letter in the space provided.

1. Eclipse __________
2. Shadow __________
3. Solar Eclipse __________
4. Lunar Eclipse __________
5. Umbra __________
6. Penumbra __________
7. Total Eclipse __________
8. Partial Eclipse __________
9. Orbit __________
10. Phase __________

A. The dark, central part of a shadow where light is completely blocked

B. The dark shape created when an object blocks light

C. When the Sun or Moon is completely covered by a shadow

D. The path that an object in space follows as it moves around another object

E. When the Moon moves between the Earth and the Sun, blocking the Sun's light

F. When only part of the Sun or Moon is covered by a shadow

G. When one object in space moves into the shadow of another object, making it hard or impossible to see the first object

H. The lighter, outer part of a shadow where only part of the light is blocked

I. A stage in the cycle of the Moon or Sun as seen from Earth

J. When the Earth moves between the Sun and the Moon, casting a shadow on the Moon

Week Three

Day 5

Compare and contrast the visual appearance of a total solar eclipse and a total lunar eclipse. Describe at least one similarity and one difference.

Week Four

Life CYCLE of STARS

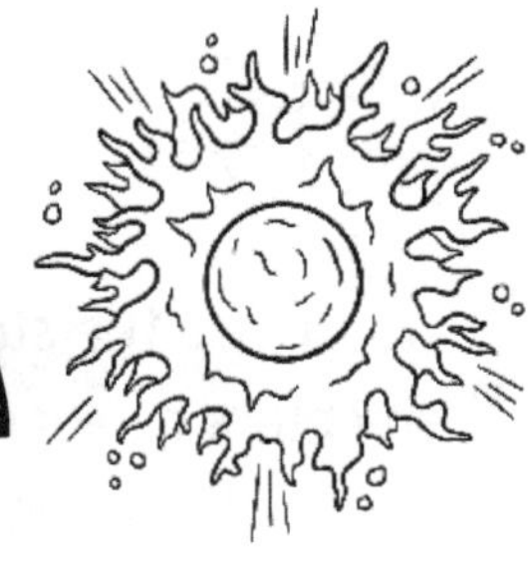

When a star like our Sun runs out of fuel, it puffs up into a red giant, expanding so much that it could swallow planets in its path.

Week Four

A CLOSER LOOK AT STARS

Read the passage to complete the tasks for the next few days.

Did you know that there are trillions of stars in the sky? We can't see most of them without a special telescope because of how far away they are. Even with a telescope, we might not be able to see all of the stars in the universe. The universe is so large! The Sun is the closest star to Earth.

Stars are giant balls of hot, glowing gas made mostly of hydrogen and helium. Stars get very hot because of the intense pressure and heat at their cores. This leads to **nuclear fusion**. Nuclear fusion is when hydrogen combines to form a gas called helium, releasing a massive amount of energy. This energy is responsible for a star's heat and brightness..

The Life Cycle of a Star

A small star goes through a different life cycle than a large star. A star, like our Sun, begins as a huge cloud of gas and dust known as a stellar nebula. Gravity causes the cloud to collapse, forming a protostar. A **protostar** is a baby star in the earliest stage of its formation.

Once the star's core heats up to millions of degrees, nuclear fusion begins. The star then enters a long phase called the **main sequence**. This is the phase in which stars spend the majority of their lives. Main sequence stars constantly burn hydrogen for billions of years.

When the hydrogen in a star's core runs out, its outer layers expand and it becomes a **red giant**. In about 5 billion years, our Sun will become a red giant. The red giant then sheds its outer layers, forming a colorful cloud of gas and dust called a **planetary nebula**. The leftover core becomes a **white dwarf**, which slowly cools over time.

Larger stars may go through a different process and become red supergiants. These stars eventually explode in a **supernova**. Depending on how large the star is, what remains can become a neutron star or a black hole.

Week Four

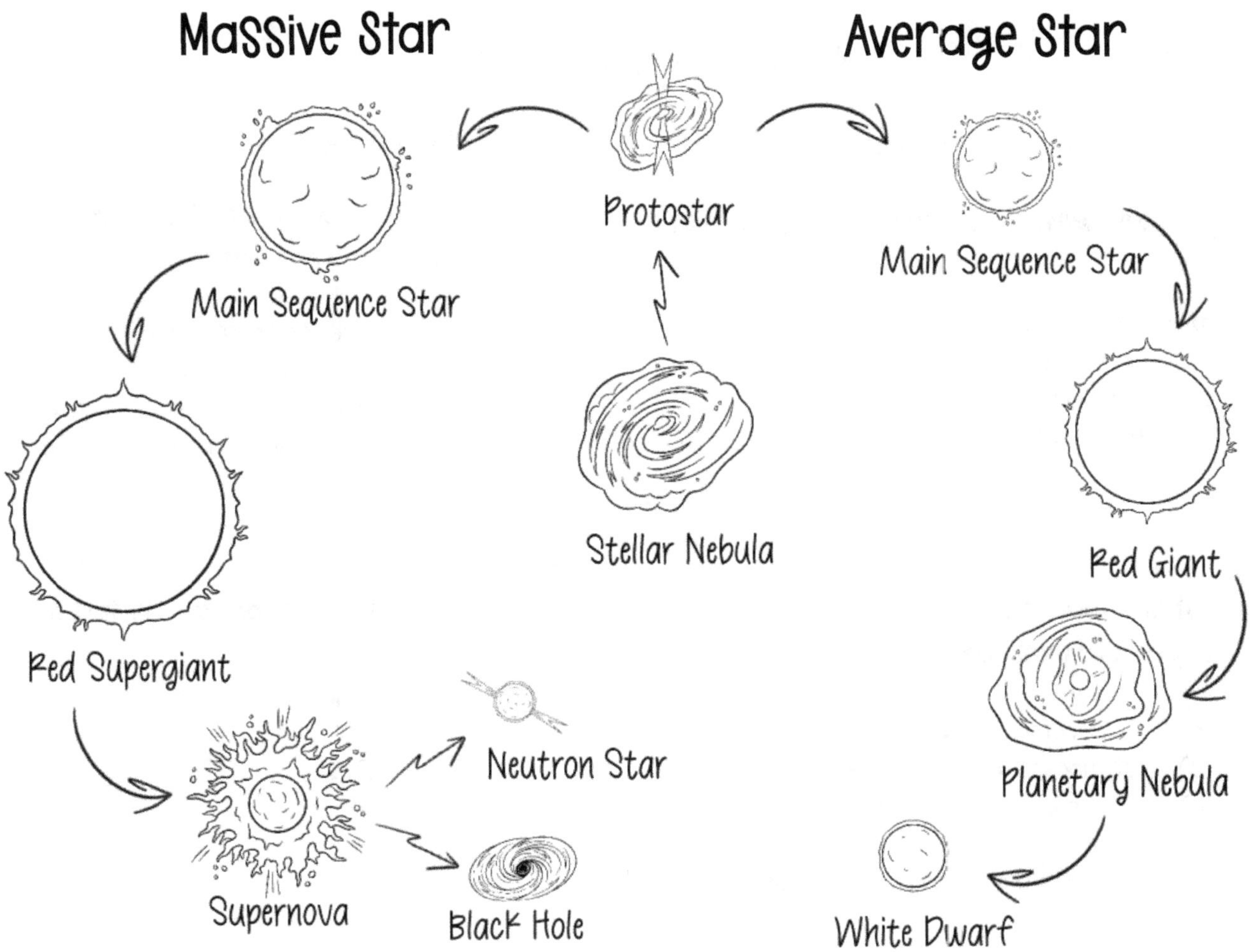

The Colors of Stars

Stars in the main sequence phase are categorized by their color. A star's color tells us its temperature. Blue stars are the hottest, while red stars are cooler. From hottest to coolest, star colors are blue, white, yellow, orange, and red.

The larger a main sequence star is, the hotter and brighter it is. Smaller stars are red and not very bright. Medium-sized stars are yellow with a medium temperature, like our Sun. The largest stars are blue and extremely hot and bright. Blue stars also burn out the fastest.

Week Four

Day 1

Name: ______________________

A CLOSER LOOK AT STARS

After learning about the stars, answer the following questions.

1. **What elements are stars mostly made of?**

   a) Oxygen and Carbon
   b) Iron and Nickel
   c) Nitrogen and Argon
   d) Hydrogen and Helium

2. **What happens to a star when it becomes a red giant?**

   a) It shrinks and cools down
   b) Its outer layer expands
   c) It turns into a black hole immediately
   d) It stays the same size but gets hotter

3. **What happens to a star after it goes through the supernova explosion?**

   a) It becomes a red giant
   b) It turns into a white dwarf
   c) It can become either a neutron star or a black hole
   d) It starts the main sequence phase again

4. **What color are the hottest stars?**

   a) Blue
   b) Yellow
   c) Red
   d) White

5. **Describe the difference between a massive star and an average star as they go through their life cycles. Include what each becomes at the end of its life.**

Name: ______________________________

Week Four

Day 2

THE LIFE CYCLE OF A MASSIVE STAR

Number the stages of a massive star in the correct order from 1 to 6, with 1 being the first stage and 6 being the last stage.

PROTOSTAR	RED SUPERGIANT	NEUTRON STAR OR BLACK HOLE
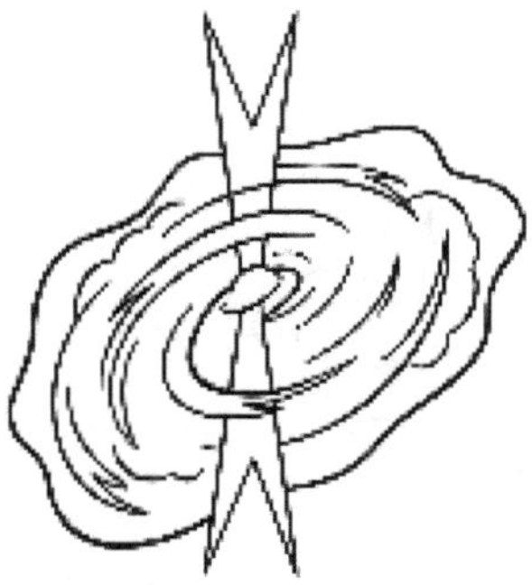		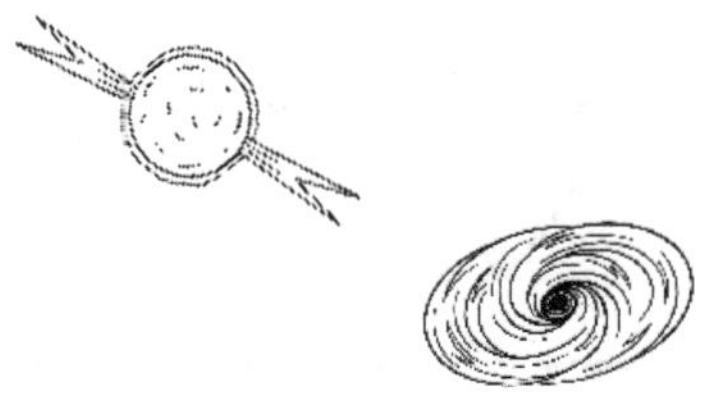
☐	☐	☐
SUPERNOVA	**STELLAR NEBULA**	**MAIN SEQUENCE STAR**

☐	☐	☐

Name: ______________________________

CREATE A STAR LIFE CYCLE COMIC STRIP

Think about the different stages in the life cycle of a star. Title each stage and write a caption to describe what is happening in each panel. Explain the scientific processes in a fun way. Two panels are already completed for you. Add details to the illustrations, such as faces or space elements, to make your comic strip more engaging. Then color or decorate your comic strip.

Name: ______________________

VOCAB MATCH

Match each word with its definition. Write the correct letter in the space provided.

1. Nuclear Fusion ________
2. Stellar Nebula ________
3. Protostar ________
4. Main Sequence ________
5. Red Giant ________
6. Planetary Nebula ________
7. White Dwarf ________
8. Black Hole ________
9. Red Supergiant ________
10. Supernova ________
11. Neutron Star ________

A. A young, forming star in the early stage of its life

B. A very dense core left behind after a supernova explosion

C. A huge cloud of gas and dust where stars begin to form

D. A very large, massive star that expands before exploding

E. A star the size of the Sun that expands as it runs out of hydrogen

F. The long stage in a star's life when it produces energy through nuclear fusion

G. The small, hot leftover core of a star after the planetary nebula stage

H. A powerful explosion that occurs when a massive star runs out of fuel

I. Hydrogen atoms combine to form helium, releasing a huge amount of energy

J. A colorful cloud of gas and dust released when a red giant sheds its outer layers

K. An object with extremely strong gravity formed when a massive star collapses completely

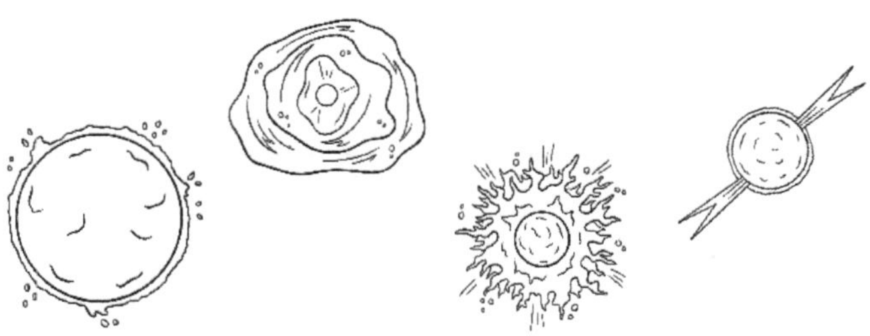

Week Four

Day 5

Write a letter to a friend explaining what happens to a star like our Sun as it grows older and becomes a red giant. Include details about how this change affects the planets around it and explain why studying the life cycle of stars is important for understanding the universe.

Week Five

CONSTELLATIONS

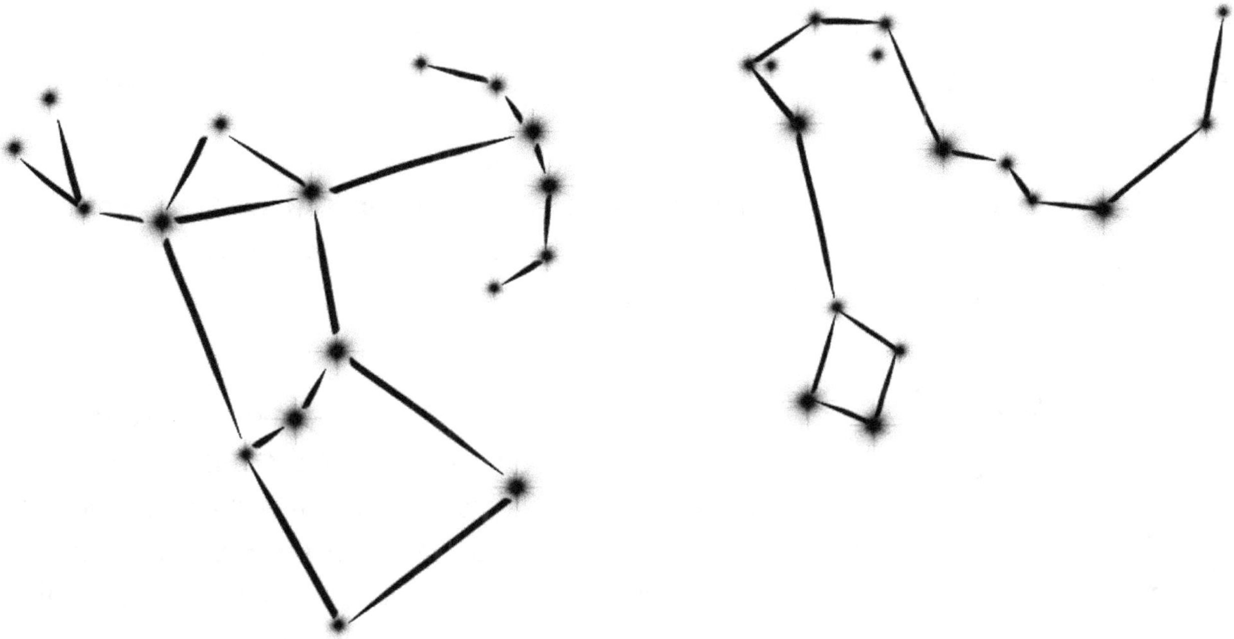

Constellations are like connect-the-dots in the night sky, forming patterns that ancient people imagined as animals, heroes, and mythical creatures!

Week Five

CONSTELLATIONS

Read the passage to complete the tasks for the next few days.

Have you ever looked up at the night sky and noticed patterns formed by the stars? These patterns can look like people, animals, or objects. Groups of stars that form imaginary shapes are called **constellations**.

The people of ancient Greece were among the first to describe constellations. They created stories, called **myths**, to explain the images they saw in the sky. For example, **Draco** is a constellation known as the Dragon. In Greek mythology, Draco guarded golden apples that belonged to Hera, the queen of the gods. Today, Draco is recognized as one of the 88 officially named constellations.

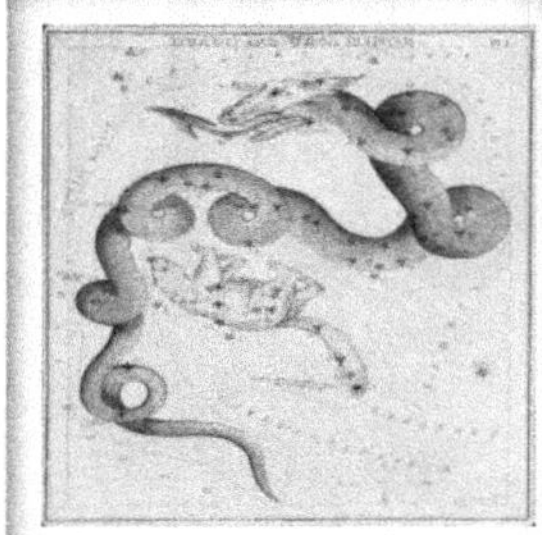

Draco & Ursa Minor Constellations

People used the stars to help them navigate, tell time, and understand the changing seasons. Farmers relied on constellations to know when to plant and harvest crops. Sailors used the stars to travel across oceans and explore new lands.

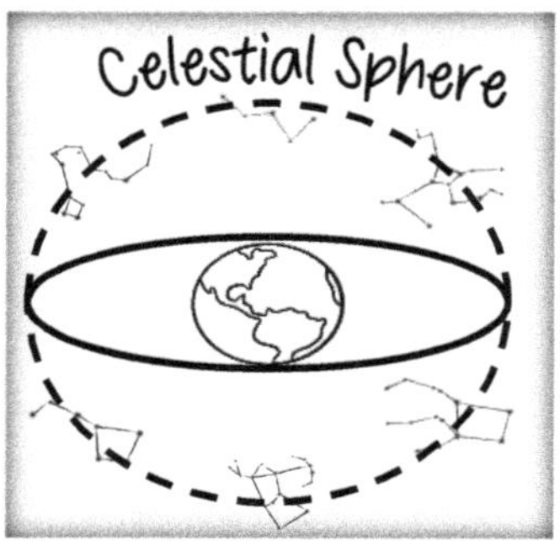

Constellations appear to change position in the sky throughout the night and during different seasons of the year. This happens because Earth is constantly moving. Astronomers use an imaginary model called the **celestial sphere** to map the positions and movements of stars and other objects in the sky. This model helps explain where constellations appear in the sky at different times.

In ancient times, people believed that stars were attached to the celestial sphere and that the entire sky revolved around Earth. We now know this is not true. The apparent movement of stars from east to west is caused by Earth rotating on its axis. While stars do have their own motion in space, they are so far away that these movements are usually not noticeable to the naked eye.

Name: ____________________

CONSTELLATIONS

After learning about the constellations, answer the following questions.

1. What are constellations?

a) Stars that are closer to the Earth
b) Groups of stars forming an imaginary shape
c) Planets in our solar system
d) Satellites orbiting Earth

2. What does the word "myth" mean?

a) Traditional story
b) Scientific explanation
c) Recent discovery
d) Fictional novel

3. Why did early civilizations use constellations?

a) To tell stories
b) To predict weather changes
c) To communicate with other civilizations
d) To know where they were and what time of year it was

4. What is the celestial sphere?

a) A real sphere in space
b) A planet in our solar system
c) An imaginary sphere used by astronomers
d) A new type of star

5. What is the myth associated with the constellation Draco?

a) A giant bear
b) A dragon that guarded the golden apples
c) A lion that ruled the sky
d) A star that lived in the ocean

6. True or False: In ancient times, people believed that stars were attached to the celestial sphere and revolved around the Earth.

7. What causes the apparent motion of stars from East to West in the night sky?

a) The Earth is rotating on its axis
b) The stars are actually moving in that direction
c) The Sun's light pushes the stars
d) The wind blows them across the sky

8. Why are the movements of stars unnoticeable to the naked eye?

a) Because stars do not move at all
b) Because of the Earth's gravity holding them in place
c) Due to their great distance from Earth
d) Because they are hidden behind clouds

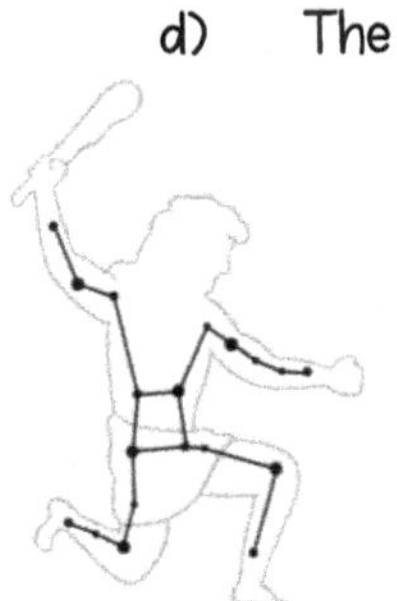

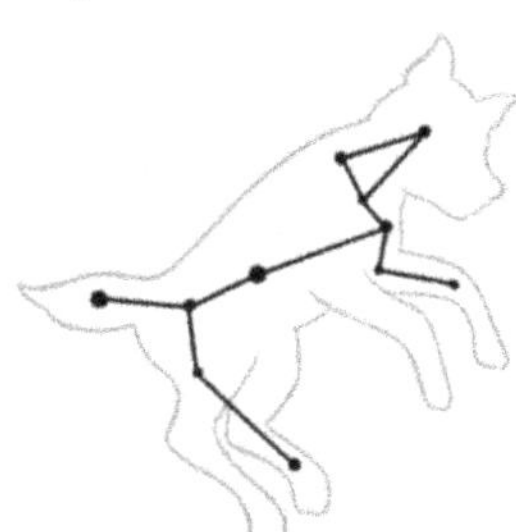

Week Five

Day 2

Name: ______________________________

CREATE A CONSTELLATION!

Connect 6 or more stars to form a constellation of anything you want: It could be an animal, a person, or an object.

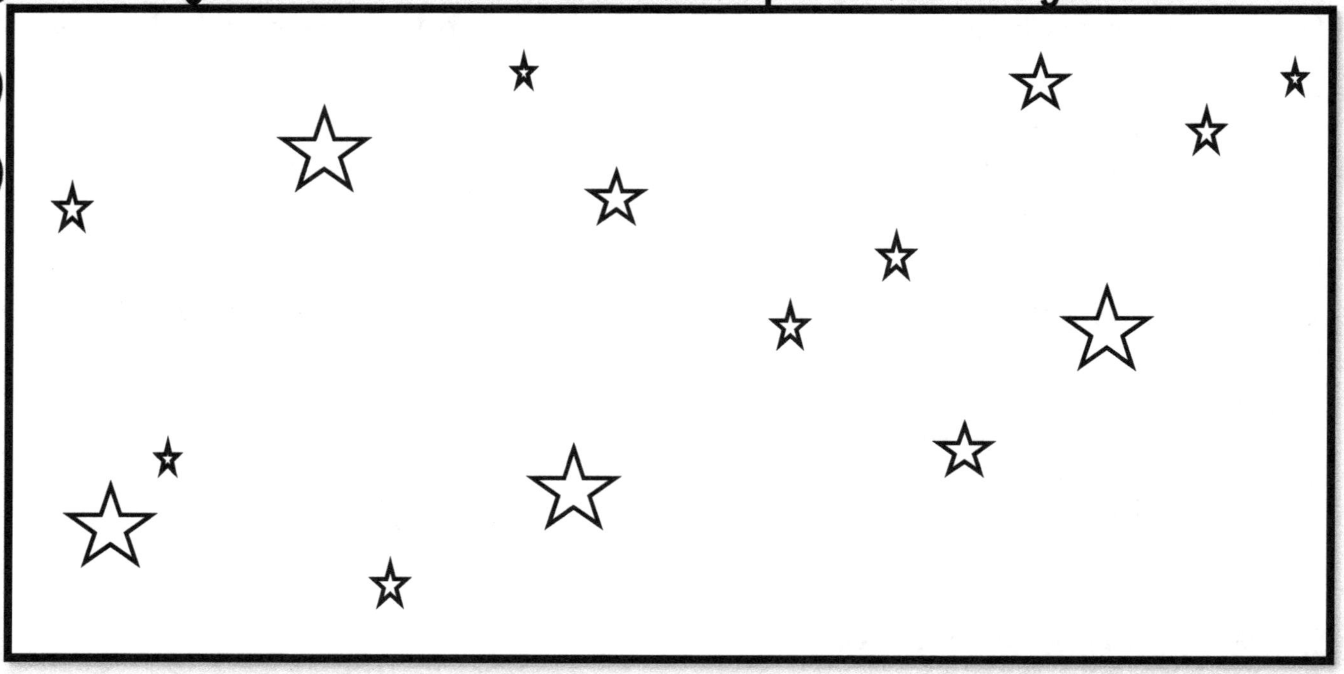

What do you want to name your constellation?

Name: ______________________________

Also Known As: ______________________________

Write a short story about the constellation you designed.

____

____

____

____

____

____

Name: ______________________

CONSTELLATION MYTHS

Orion

Orion was a skilled hunter, famous for his strength and beauty. According to Greek mythology, he bragged that he could kill any creature on Earth. This angered the Earth goddess Gaia, who sent a giant scorpion to sting him. After his death, the gods placed him in the sky as a constellation.

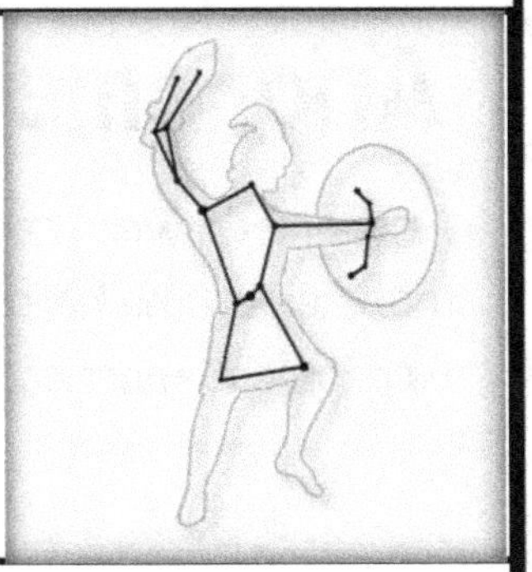

Ursa Major

Ursa Major, also known as The Great Bear, includes the famous group of stars called the Big Dipper. According to myth, Callisto, a friend of the goddess Artemis, was transformed into a bear by the jealous Hera. To protect her, Zeus placed her in the sky, where she became Ursa Major.

Draco

Draco, the dragon, guarded the golden apples in the garden of Hesperides. These apples were a wedding gift to Hera, the queen of the gods. Draco was eventually slain by Hercules as one of his 12 labors, and he was placed in the sky as a constellation.

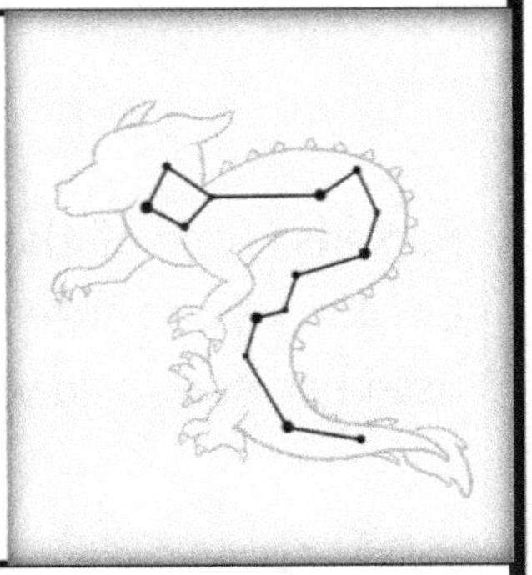

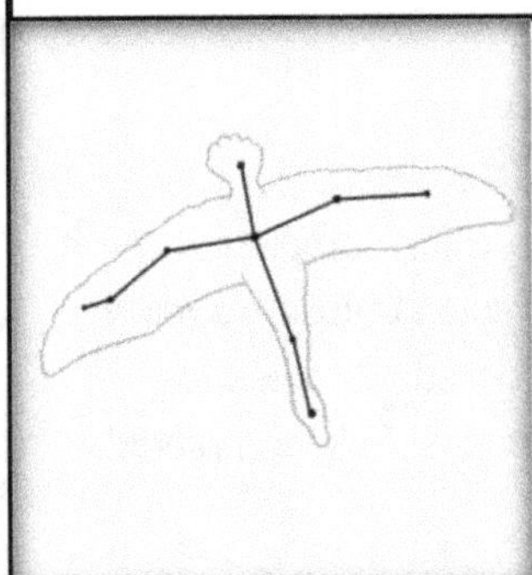

Cygnus

Cygnus is also known as 'The Swan'. One of the most famous myths associated with Cygnus involves the character Leda, Queen of Sparta. Zeus liked Leda. One day, Leda was attacked by an eagle, so he disguised himself as a swan and rescued her.

Cassiopeia

Cassiopeia is also known as 'The Vain Queen'. She bragged that she and her daughter Andromeda were more beautiful than the sea nymphs. This angered the god of the sea, Poseidon, who sent a monster to destroy her kingdom. To save her people, Cassiopeia had to sacrifice her daughter Andromeda. However, Perseus rescued Andromeda by defeating the sea monster. The gods later placed Cassiopeia in the sky as punishment.

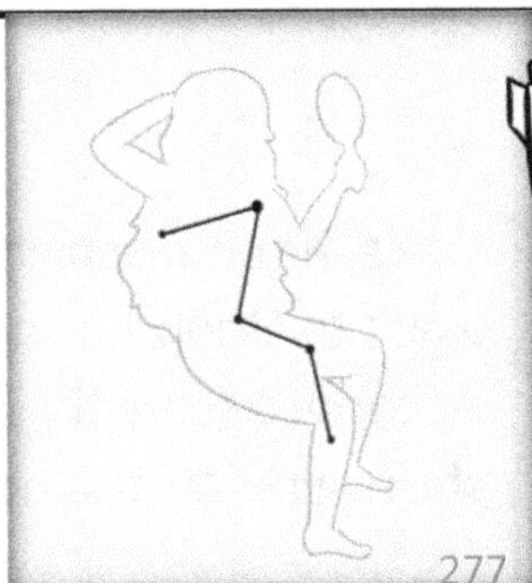

Name: ____________________

Week Five

Day 3

CONSTELLATION MYTHS

Andromeda

Andromeda was the daughter of Cassiopeia and was saved from a sea monster by the hero Perseus. After Perseus killed the monster, he and Andromeda married, and she was placed in the sky as a constellation.

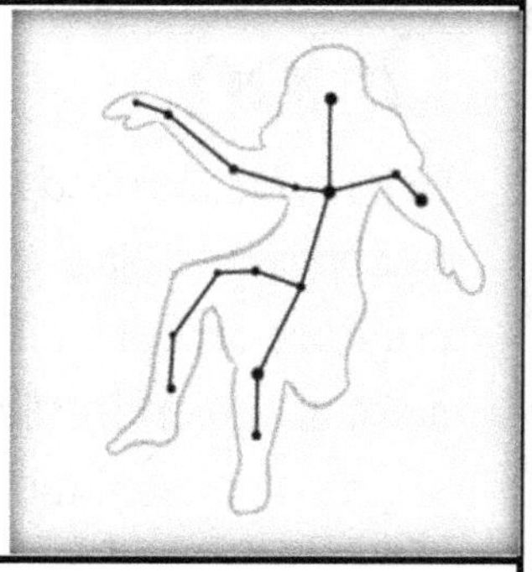

Pegasus

Pegasus is a winged horse born from the blood of Medusa after she was killed by Perseus. Pegasus flew to Mount Olympus and carried thunderbolts for Zeus. He also helped the hero Bellerophon, but Zeus later sent a gadfly to stop Bellerophon when he tried to fly to Olympus.

Perseus

Perseus is one of the greatest Greek heroes, known for killing Medusa, the Gorgon whose gaze could turn people to stone. After defeating Medusa, he used her head as a weapon to save Andromeda from a sea monster.

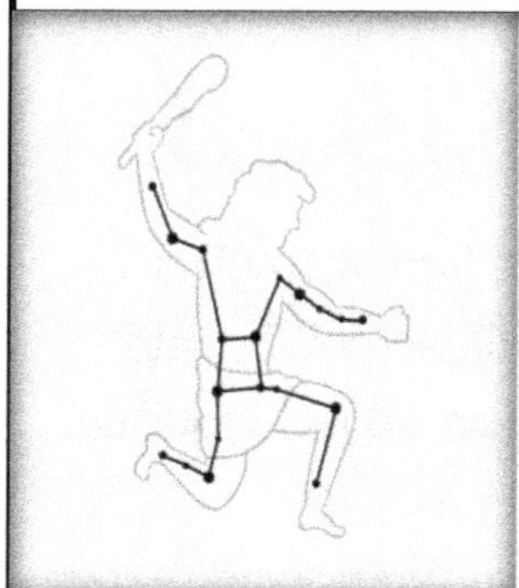

Hercules

Hercules, also known as the strongman, is a famous hero known for completing the twelve nearly impossible labors. The goddess Hera caused Hercules to lose his mind and do terrible things. As punishment, the King Eurystheus, ordered Hercules to complete the 12 extremely difficult and dangerous tasks. He successfully completed them all, showing his incredible strength and bravery.

Canis Major

Canis Major, meaning "Greater Dog", represents the dog Laelaps, who was a gift to Orion. Laelaps was known for his incredible speed and loyalty. According to myth, Zeus placed him in the sky as a constellation to run eternally after the hare (represented by the constellation Lepus).

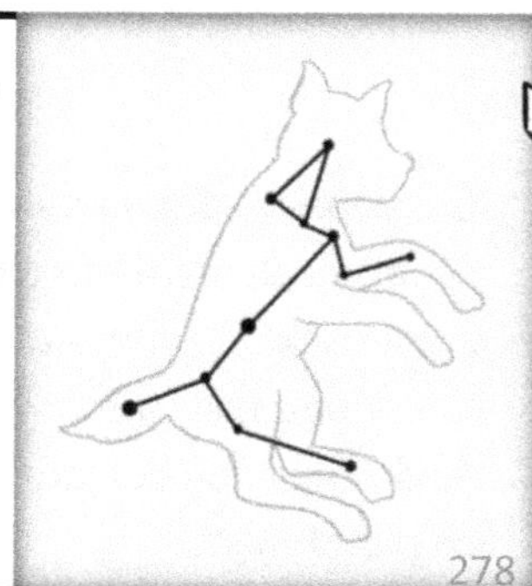

Week Five

Day 3

Name: ______________________________

WHO AM I? (CONSTELLATIONS)

#	Clue		Constellation
1	I was going on adventures with Bellerophon until I got stung by a gadfly.		Draco 
2	I am also known as 'The Big Dipper'.		Andromeda 
3	I was saved by Perseus from a sea monster.		Perseus 
4	I was a gift to Orion and am known for my speed and loyalty.		Orion 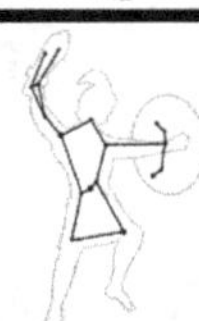
5	I guarded the golden apples of Hesperides.		Cygnus 
6	I was known for the twelve labors.		Canis Major 
7	I bragged that my daughter was more beautiful than the sea nymphs.		Ursa Major 
8	I was known for killing Medusa.		Hercules 
9	I am a skilled hunter who was stung by a scorpion.		Pegasus 
10	I was placed among the stars as a swan.		Cassiopeia 

Name: ______________________

Week Five

Day 4

VOCAB MATCH

Match each word with its definition. Write the correct letter in the space provided.

1. Constellation ________
2. Astrologer ________
3. Cluster ________
4. Star ________
5. Myth ________
6. Celestial Sphere ________

A. A group of stars that are close together in space

B. An imaginary sphere that is used by astronomers to explain the positions and movements of stars and other objects

C. A person who studies the positions of stars and planets and believes they influence personality and behavior

D. A massive ball of gas that emits light and heat, often forming part of a constellation

E. A group of stars that form a pattern in the sky

F. A traditional story that a group of people believed to be real

Also Known As...

WORD BANK: The Winged Horse | The Captive Princess | The Swan

7. 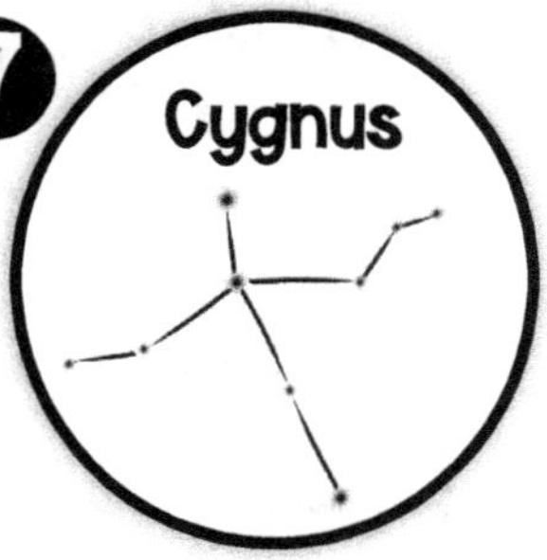

____

8. 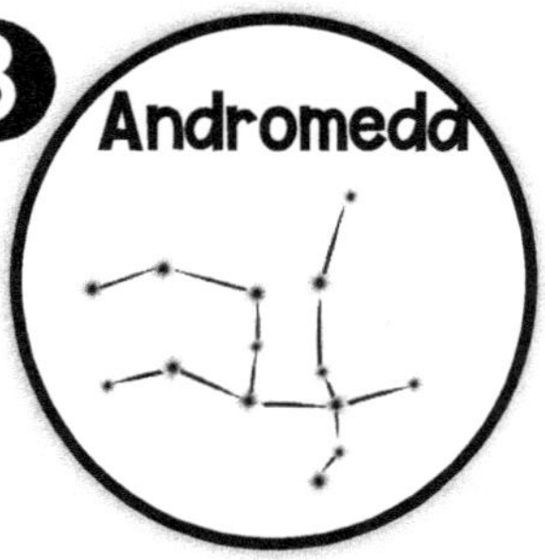

____

9. 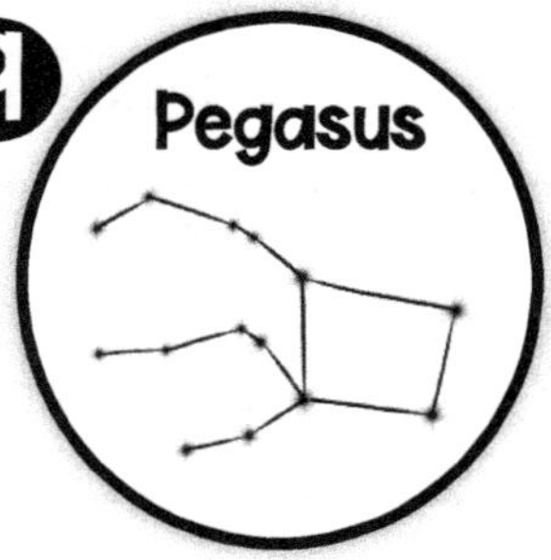

____

Week Five

Day 5

Different cultures around the world have different names and stories for the same constellations. Why do you think this happens? What does this tell us about how people view the night sky?

Week Six

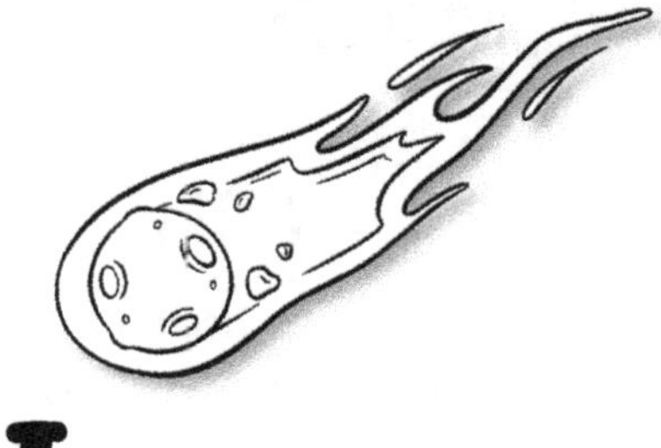

Comets, METEORS, and ASTEROIDS

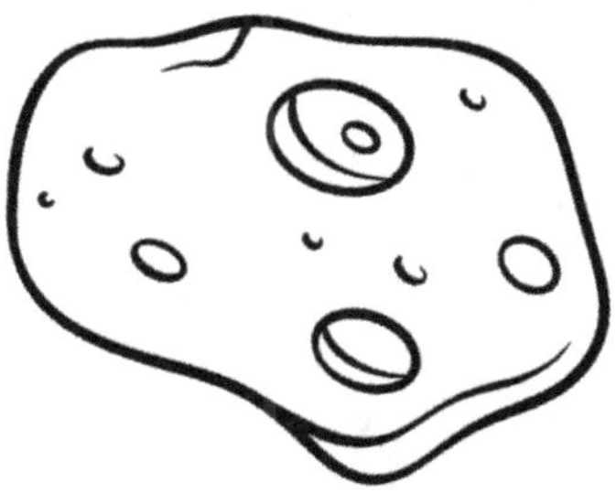

Comets, meteors, and asteroids all have this thing in common—they're fascinating space rocks that give us a glimpse of the wonders of our solar system!

Week Six

COMETS, METEORS, AND ASTEROIDS

Read the passage to complete the tasks for the next few days.

Did you know that there are many kinds of "space rocks" zooming through our solar system? Three of the most interesting ones are comets, meteors, and asteroids. While they may seem similar, each one has unique features that make it different from the others.

COMETS

Comets are often called "dirty snowballs" because their solid center, called the **nucleus**, is made of ice, dust, and rock. As a comet travels closer to the Sun, heat causes the ice in the nucleus to turn into gas. This forms a bright, glowing "head" around the comet called a **coma**. As the comet moves through space, gas and dust trail behind it, creating two tails: a gas tail and a dust tail. The **gas tail** always points directly away from the Sun because of the solar wind, while the **dust tail** curves slightly along the comet's path. Sometimes, comets can be seen in the night sky as they pass near Earth, appearing as bright objects with long, glowing tails.

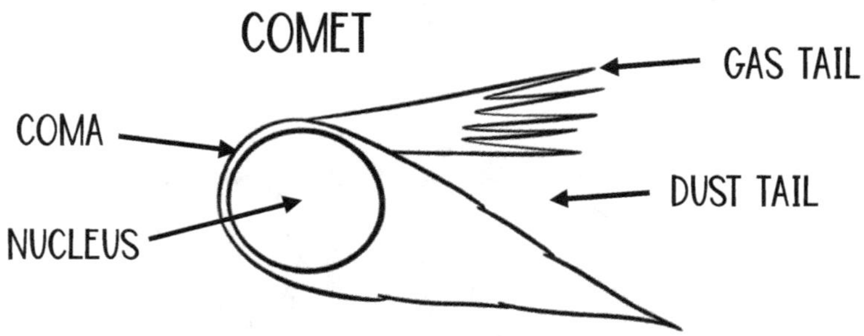

METEORS

Meteors are often called "shooting stars," but they are not actually stars at all. Before entering Earth's atmosphere, these small pieces of rock or dust are called **meteoroids**. When a meteoroid enters the atmosphere, it travels extremely fast and burns up, creating a streak of light in the sky. This streak of light is called a **meteor**. When many meteors appear in the sky at the same time and in the same area, the event is called a **meteor shower**. Most meteors burn up completely before reaching the ground. However, if a piece survives the trip through the atmosphere and lands on Earth's surface, it is called a **meteorite**.

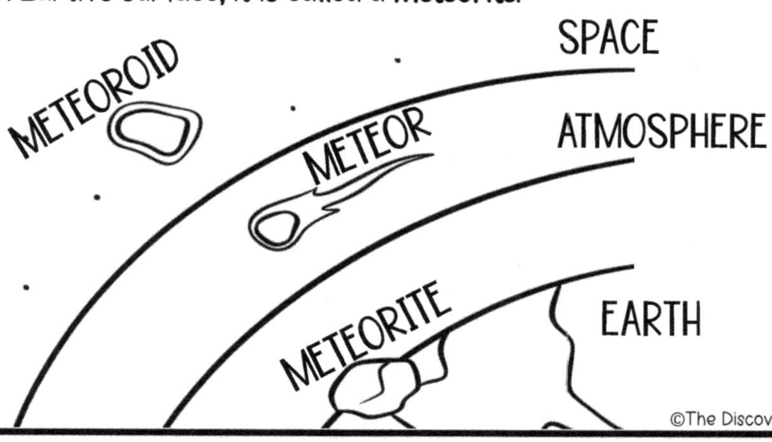

ASTEROIDS

Asteroids are large chunks of rock that orbit the Sun, similar to planets. Most asteroids are found in the **asteroid belt**, a region of space located between Mars and Jupiter. There are millions of asteroids in this belt, and they come in many different sizes. Some are as small as pebbles, while others can be hundreds of miles wide..

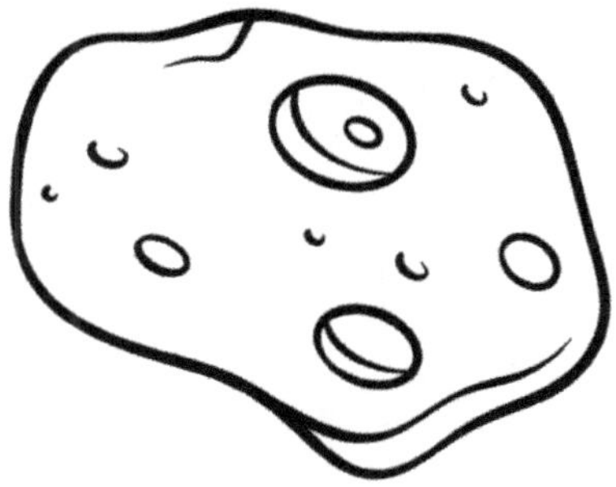

Scientists continue to study comets, meteors, and asteroids to learn more about the history of our solar system. Every year, new space rocks are discovered, and each one helps scientists better understand the universe. Space is full of mysteries, and there is always more to explore. Who knows? One day, you might be the person who makes the next big discovery in space!

Name: ______________________

Week Six

Day 1

COMETS, METEORS, AND ASTEROIDS

After learning about the comets, meteors, and asteroids, answer the following questions.

1. **What is a comet mostly made of?**

   a) Metal and rock
   b) Ice, dust, and rocks
   c) Water and soil
   d) Only gases

2. **What happens to a meteoroid when it enters Earth's atmosphere?**

   a) It burns up and creates a streak of light
   b) It turns into an asteroid
   c) It grows larger
   d) It freezes

3. **Where are most asteroids found in our solar system?**

   a) Between Earth and Mars
   b) Near the Sun
   c) Outside the solar system
   d) Between Mars and Jupiter

4. **What is a meteorite?**

   a) A meteor that burns up in the atmosphere
   b) A small asteroid orbiting the Sun
   c) A meteoroid that reaches Earth's surface
   d) A comet without a tail

5. **Meteor showers happen when many meteors appear in the sky at the same time. Have you ever seen a meteor shower? If so, describe what it was like. If not, would you like to see one? Why do you think people find meteor showers so fascinating?**

Week Six

Day 2

Name: ______________________________

COMET, METEOR, OR ASTEROID?

Read each statement carefully. Decide whether it describes a comet, meteor, or asteroid. Place a check mark in the correct column.

	COMET	METEOR	ASTEROID
1. Known for having a bright, glowing tail that can be seen from Earth			
2. A small, rocky body that orbits the Sun, often found in a belt between Mars and Jupiter			
3. Often called a "shooting star" when it burns up in Earth's atmosphere			
4. If this object reaches Earth's surface without burning up, it is called a meteorite.			
5. This object can be as small as a pebble and can also be hundreds of miles wide			
6. Mostly made of ice, dust, and rocks			
7. Can be seen as a streak of light in the night sky when it burns up in the atmosphere			
8. Develops a coma as it nears the Sun			

Name: ______________________

Week Six

Day 3

FILL IN THE BLANKS

Fill in the blanks with the correct words. Then, find your answers in the word search on the next page.

1. ______________ are often called "dirty snowballs" because they are made of ice, dust, and rocks.
2. When a comet gets close to the Sun, it develops a glowing head called a ______________.
3. The tail of a comet always points ______________ from the Sun due to the solar wind.
4. If a meteor survives its journey through the atmosphere and reaches Earth's surface, it is called a ______________.
5. The ______________ Belt, located between Mars and Jupiter, contains millions of rocky objects.
6. Asteroids ______________ the Sun just like planets.
7. Comets have two tails: a gas tail and a ______________ tail, both of which become visible as they approach the Sun.
8. Most ______________ are small and burn up completely before they reach the Earth's surface.
9. A meteor ______________ occurs when many meteors appear in the sky at the same time and from the same direction.
10. Scientists around the world continue to study space rocks to learn more about the ______________ of our solar system.

Name: ______________________

FIND YOUR ANSWERS

Find the answers you wrote on the previous page in the word search.

C	E	C	A	D	U	E	N	E	C	O	M	A	T
A	S	E	P	S	O	P	E	I	T	I	M	S	G
M	H	A	B	D	T	U	M	A	S	I	I	N	B
O	O	W	D	C	M	E	T	E	O	R	S	L	E
U	W	L	M	A	C	S	R	O	N	E	R	G	H
F	E	A	S	R	P	O	Y	O	I	C	A	L	A
L	R	O	Y	S	R	T	M	R	I	I	T	N	V
A	I	S	P	Y	E	O	E	E	T	D	I	S	I
H	O	R	A	M	D	L	T	T	T	O	O	F	O
E	I	W	E	E	A	B	E	E	I	S	N	S	R
Y	A	S	E	R	T	I	O	O	R	O	F	I	T
D	S	B	T	Y	O	V	R	U	Z	Z	N	I	L
U	H	I	B	O	R	N	I	T	I	O	B	Q	N
S	R	A	V	F	R	Y	T	U	V	R	G	Z	L
T	H	O	J	E	R	Y	E	N	O	A	D	F	R
E	H	K	L	I	U	O	F	X	L	O	B	Y	I

Name: ____________________

Week Six

Day 4

Vocab Match

Match each word with its definition. Write the correct letter in the space provided.

1. Asteroid ______
2. Comet ______
3. Meteor ______
4. Meteorite ______
5. Meteoroid ______
6. Coma ______
7. Nucleus ______
8. Asteroid Belt ______
9. Solar Wind ______
10. Orbit ______

A. A streak of light in the sky, often called a "shooting star," caused by a meteoroid burning up in Earth's atmosphere

B. The glowing head of a comet, formed when its ice turns into gas as it gets closer to the Sun

C. A "dirty snowball" made of ice, dust, and rocks that orbits the Sun and develops a glowing tail when near the Sun

D. The path an object takes as it moves around a star, planet, or moon

E. A large rock in space that orbits the Sun, mostly found in the asteroid belt

F. A region in space between Mars and Jupiter where most asteroids are found

G. A piece of a meteoroid that survives its trip through Earth's atmosphere and lands on the ground

H. The solid center of a comet, made of ice and rock

I. A stream of charged particles released from the Sun that pushes a comet's gas tail away from the Sun

J. A small rock or particle in space, smaller than an asteroid

Week Six

Day 5

Write a letter to a friend explaining the differences between an asteroid, a meteor, and a comet. Use examples to describe how each one moves and behaves in space.

Week Seven

GRAVITY

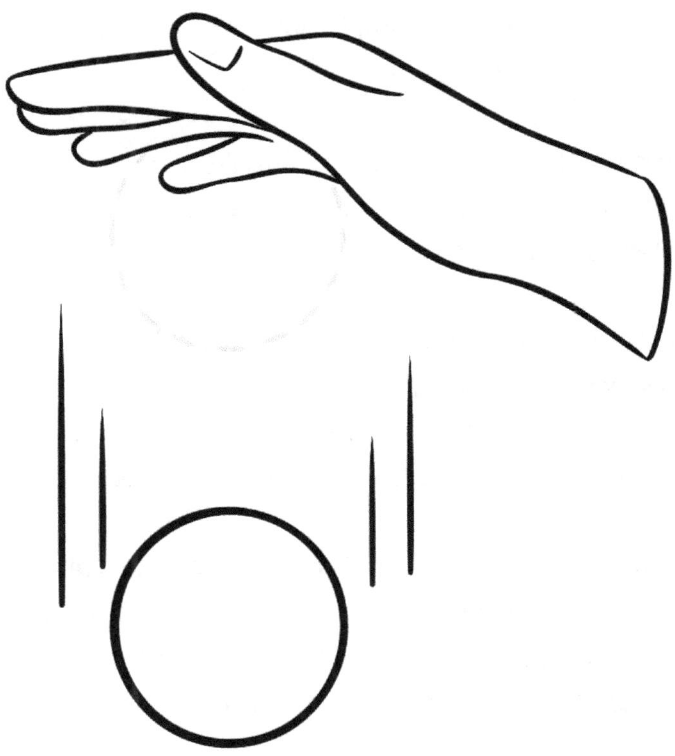

Gravity is like an invisible glue that keeps us and everything else stuck to the Earth, preventing us from floating away into space!

GRAVITY

Week Seven

UNDERSTANDING GRAVITY IN SPACE

Read the passage to complete the tasks for the next few days.

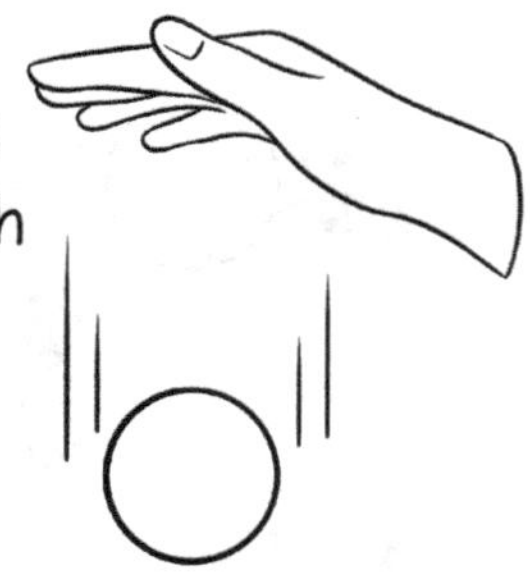

Have you ever wondered why you don't keep floating off into space when you jump? The answer lies in understanding gravity. Gravity is a force that pulls objects toward one another. It is what keeps us on Earth instead of drifting away into space. On Earth, gravity pulls everything downward toward the center of the planet, which is why objects fall when you let go of them.

In space, gravity plays a huge role in how everything moves. The Sun, the largest object in the solar system, has the strongest gravity. Its gravity is so powerful that it pulls all the planets, including Earth, into orbit around it, keeping them from flying off into space.

The Moon also has gravity, which is why it stays in orbit around Earth. The Moon's gravity even affects the tides in our oceans. When the Moon is closer to one side of Earth, it pulls on the water, causing high tides.

Every object in space has some gravity, even if it is very small. Larger objects, such as planets and stars, have stronger gravitational forces because they have more mass. Mass is the amount of matter an object has. The more mass an object has, the stronger its gravitational pull. Because of this, smaller objects are pulled toward objects with much greater mass.

Gravity pulls all objects toward Earth at the same rate, regardless of their mass. This means that if you were to drop a feather and a rock at the same time in a vacuum chamber (where there is no air), they would hit the ground at the same time. However, air resistance on Earth causes lighter objects, like feathers, to fall more slowly.

Gravity also becomes weaker the farther you are from an object. That is why you would weigh less on the Moon than you do on Earth. The Moon has less mass and therefore less gravity.

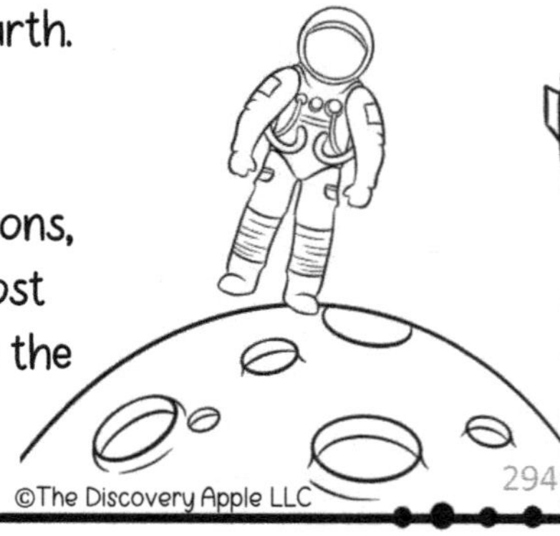

Without gravity, there would be no orbits, and planets, moons, and even galaxies would not stay in place. Gravity is one of the most important forces in the universe, making sure everything moves in the right way and stays where it is supposed to be.

Name: ______________________________

Week Seven

Day 1

UNDERSTANDING GRAVITY IN SPACE

After learning about the gravity, answer the following questions.

1. **What is gravity?**

   a) A force that pushes objects away from each other
   b) A force that pulls objects toward each other
   c) A force that only affects large objects
   d) A force that only exists in space

2. **Why do objects fall to the ground when you let go of them on Earth?**

   a) Because of the wind
   b) Because of air resistance
   c) Because of gravity
   d) Because they have mass

3. **Why does the moon stay in orbit around the Earth?**

   a) Because of Earth's gravity
   b) Because it has not gravity
   c) Because of its speed
   d) Because its close to the sun

4. **What happens to objects when they move farther away from a large object with strong gravity?**

   a) They weigh more
   b) They have the same weight
   c) They disappear
   d) They weigh less

5. **Explain why gravity is important in the solar system. How does it affect the movement of planets?**

Name: ______________________

GRAVITATIONAL PULL

Number the objects from 1 to 5. Write 1 next to the object with the greatest gravitational pull and 5 next to the object with the least gravitational pull.

THE MOON

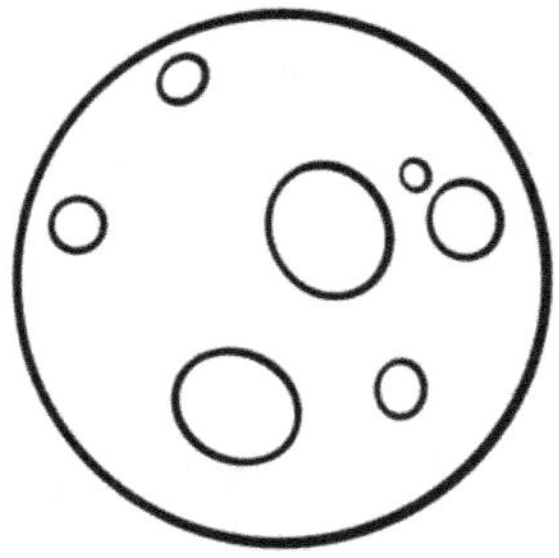

A CAR

THE SUN

AN ASTEROID

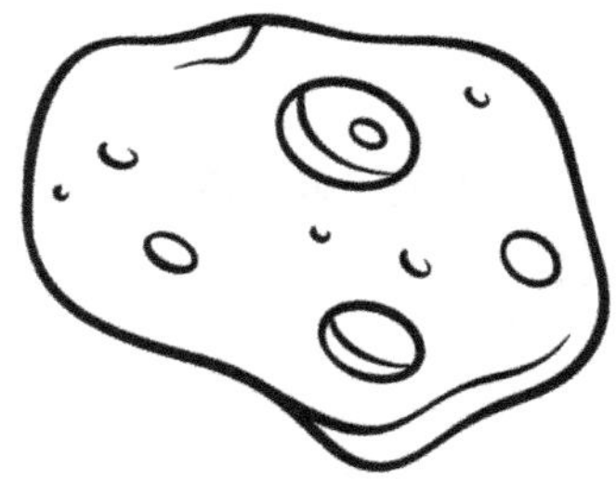

A BASEBALL

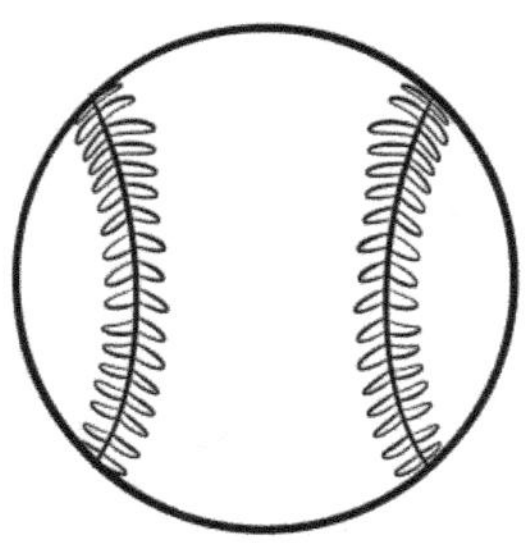

EARTH

GRAVITY

Name: ____________________

Week Seven

Day 3

TRUE OR FALSE

Mark each statement as true or false.

Extension: For each false statement, rewrite the sentence to make it true on a separate sheet of paper.

		True	False
1.	Gravity is a force that pushes objects away from each other.	☐	☐
2.	Gravity is what keeps the planets in orbit around the Sun.	☐	☐
3.	Gravity affects the tides in Earth's oceans by pulling on the water.	☐	☐
4.	Gravity only affects objects that are in motion.	☐	☐
5.	The Moon's gravity is weaker than Earth's gravity.	☐	☐
6.	An object's weight and mass are the same on every planet.	☐	☐
7.	Gravity only exists on planets and not in space.	☐	☐
8.	Earth's gravity is what causes objects to fall toward the ground when dropped.	☐	☐
9.	Without gravity, objects in space would eventually stop moving.	☐	☐
10.	The gravitational force between two objects decreases as the distance between them increases.	☐	☐

GRAVITY

Name: ____________________

Week Seven

Day 4

Vocab Match

Match each word with its definition. Write the correct letter in the space provided.

1. Gravity ________
2. Mass ________
3. Weight ________
4. Acceleration ________
5. Air Resistance ________
6. Tides ________
7. Force ________
8. Orbit ________
9. Microgravity ________
10. Free Fall ________

A. The rate at which an object's velocity changes due to forces like gravity

B. The amount of matter in an object, which affects how strong its gravitational pull is

C. The force of air pushing against a moving object, which can affect how it moves due to gravity

D. The curved path an object takes as it moves around another object due to gravity

E. The rise and fall of sea levels caused by the gravitational pull of the Moon and the Sun on Earth's oceans

F. The force that pulls objects with mass toward each other

G. The motion of an object falling due to gravity alone, without air resistance

H. The measure of the force of gravity on an object

I. A condition in which objects appear to be weightless and experience very weak gravitational forces, often found in space

J. A push or pull that can cause an object to move, stop, or change direction

Week Seven

Day 5

1. Explain what you think would happen if Earth's gravity suddenly became weaker? How might that also affect the way we move?

2. How does gravity help us when we throw a ball into the air? What do you think would happen if there was no gravity?

Week Eight

EARTH'S Rotation

AND Revolution

While Earth rotates to give us day and night, it also travels around the Sun in a year-long journey, completing one full revolution every 365 days!

SPINNING THROUGH SPACE

Read the passage to complete the tasks for the next few days.

Did you know that Earth is always moving? Even though you do not feel it, Earth is constantly in motion. It spins in place and travels around the Sun. The spinning of Earth is called **rotation**. Earth has an imaginary line called an **axis** that runs from the North Pole, through the center of the planet, to the South Pole. Earth spins on this axis, and it takes about 24 hours, or one day, to complete one full rotation.

As Earth rotates, different parts of the planet face the Sun at different times. When one side of Earth faces the Sun, it experiences daytime, while the other side experiences nighttime. For example, when it is daytime in the United States, it is nighttime in China because it is on the opposite side of the planet. Earth's rotation also makes the Sun appear to rise in the east and set in the west. The Sun does not actually move across the sky; this effect happens because Earth is turning.

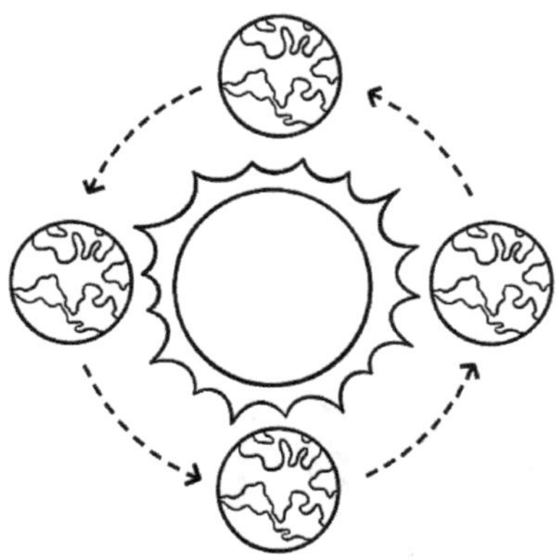

While Earth is rotating, it is also moving around the Sun. This movement is called **revolution**. Earth follows a path called an **orbit** as it moves around the Sun. Earth's orbit is not a perfect circle. It is slightly stretched out, forming an oval shape called an ellipse. It takes Earth about **365 ¼** days to complete one full revolution around the Sun. This is why one year has 365 days.

As Earth revolves around the Sun, different parts of the planet experience changing **seasons**. Earth has seasons because its axis is tilted. This tilt always points in the same direction as Earth travels around the Sun. When an area of Earth is tilted toward the Sun, it receives more direct sunlight and experiences summer. When an area is tilted away from the Sun, it receives less sunlight and experiences winter.

Week Eight

Day 1

Name: ______________________________

SPINNING THROUGH SPACE

After learning about the earth's rotation and revolution, answer the following questions.

1. How does the tilt of Earth's axis affect the seasons?

a) It causes the Earth to rotate faster in summer and slower in winter.
b) It changes the amount of sunlight each hemisphere receives, leading to different seasons.
c) It makes the Sun move closer to the Earth during summer and farther away in winter.
d) It has no effect; seasons are caused by the distance between Earth and the Sun.

2. What would happen if Earth's axis were not tilted at all?

a) The Earth would have the same season all year long
b) The days would be longer during the winter and shorter during the summer.
c) The Sun would not rise or set, staying in the same place in the sky.
d) The Earth would rotate slower, and the year would become shorter.

3. How long does it take the Earth to complete one full rotation?

a) 365 days
b) 24 hours
c) 12 hours
d) 1 month

4. Why does the Sun appear to rise in the east and set in the west?

a) Because the Sun moves around the Earth
b) Because the Earth revolves around the Sun
c) Because the Earth's axis is tilted
d) Because the Earth rotates on its axis

5. Imagine you are an astronaut observing Earth from space. How would you explain the difference between Earth's rotation and revolution to someone who has never been to Earth before?

Name: ______________________________

TRUE OR FALSE

Read each sentence and mark the appropriate box to indicate whether it is true or false.

	True	False
1. The Earth spins on its axis.	☐	☐
2. The Earth orbits around the Moon.	☐	☐
3. It takes the Earth 24 hours to do a complete spin on its axis.	☐	☐
4. The Earth spins from west to east.	☐	☐
5. The path Earth follows is called its rotation.	☐	☐
6. Revolution is the movement of Earth around the Sun.	☐	☐
7. It takes the Earth 345 ¼ days to orbit the Sun.	☐	☐
8. When the northern hemisphere is tilted away from the Sun, it experiences summer.	☐	☐
9. The top part of the Earth is called the southern hemisphere.	☐	☐
10. Seasons are caused by Earth's tilted axis as it travels around the Sun.	☐	☐

Name: ______________________________

EARTH'S ROTATION AND REVOLUTION CROSSWORD

Fill in the crossword puzzle using the clues provided. All answers are related to Earth's rotation and revolution.

ACROSS

2. - The time it takes for Earth to complete one full revolution around the Sun
5. - The path Earth follows as it revolves around the Sun
6. - The star at the center of our solar system that Earth orbits around
7. - The spinning of Earth on its axis, causing day and night
8. - The four periods of the year caused by Earth's revolution and tilted axis
10. - The angle of Earth's axis, which affects the seasons

DOWN

1. - The period of time when one part of Earth is facing the Sun
3. - Earth's movement around the Sun, taking about 365 days
4. - The period of time when one part of Earth is turned away from the Sun
9. - The imaginary line Earth spins around

Week Eight

Day 4

Name: ____________________________

VOCAB MATCH

Match each word with its definition. Write the correct letter in the space provided.

1. Rotation ________
2. Revolution ________
3. Axis ________
4. Orbit ________
5. Tilt ________
6. Hemisphere ________
7. Seasons ________
8. Equator ________
9. Elliptical ________

A. The path that an object follows as it moves around another object due to gravity

B. The oval shape of Earth's orbit around the Sun

C. The movement of one abject around another object in a circular or elliptical path

D. An imaginary line around the middle of the Earth that divides it into the Northern and Southern Hemispheres

E. Leaning on one side

F. The spinning motion of an object around its own axis

G. Half of the Earth, usually divided into Northern and Southern or Eastern and Western

H. An imaginary line running through the Earth from the North Pole to the South Pole that the Earth spins around

I. Periods of the year marked by distinct weather patterns and daylight hours, caused by Earth's tilt on its axis and its orbit around the Sun

Week Eight

Day 5

1. How would life on Earth be different if the planet did not rotate on its axis?

2. If you could design a planet with a unique rotation or orbit, what would it look like, and how would it affect the seasons or day and night?

Week Nine

The HISTORY of Space EXPLORATION

Laika's journey into space was the first time any living creature had been sent to orbit, marking a significant moment in our quest to explore beyond Earth.

Week Nine

THE HISTORY OF SPACE EXPLORATION

Read the passage to complete the tasks for the next few days.

Space exploration is an exciting journey that humans have been on for many years. It began in the 1950s, when scientists and engineers started launching rockets to learn more about space.

One of the first major events happened in 1957, when the Soviet Union, a group of countries that included Russia, launched a satellite called **Sputnik 1** into space. Sputnik 1 was the first man-made object to orbit Earth, and it amazed people around the world. It showed that humans could send objects beyond our planet.

LAIKA

A month later, the Soviet Union launched Sputnik 2, which carried a dog named Laika. **Laika** became the first animal to orbit Earth. Sadly, she did not survive the mission. Scientists believe she died a few hours after launch due to overheating. Sputnik 2 continued to orbit Earth for about five months before it eventually re-entered the atmosphere and burned up. Even though Laika's mission had a tragic ending, it provided scientists with important information about how space travel affects living beings.

Soon after, the United States joined the space race by launching Explorer 1 in 1958. **Explorer 1** was the first American satellite, and it helped scientists discover the **Van Allen radiation belts** surrounding Earth. These belts are layers of charged particles that protect our planet from harmful space radiation. After completing more than 58,000 orbits, Explorer 1 re-entered Earth's atmosphere and burned up.

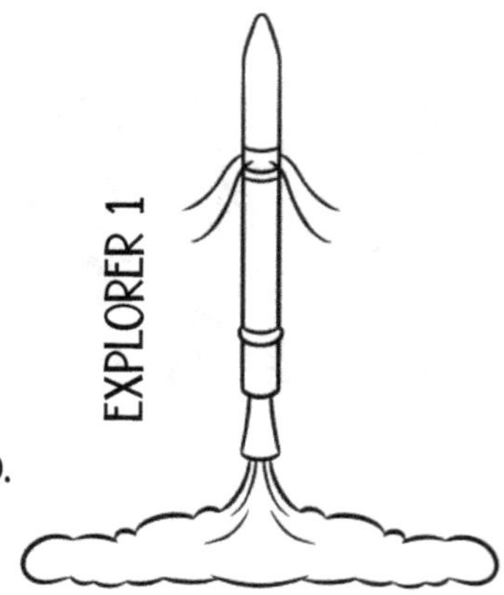

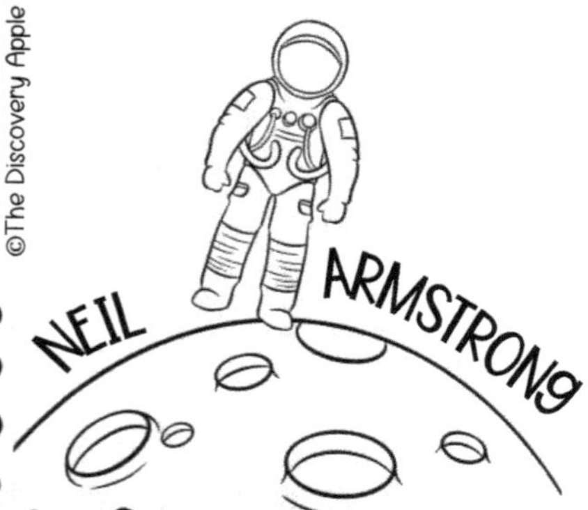

As space exploration continued, humans were ready to travel to space themselves. In 1969, the United States made history when **NASA's Apollo 11** mission successfully landed the first humans on the Moon. **Neil Armstrong** and **Buzz Aldrin** became the first people to walk on the Moon. Neil Armstrong was the first to step onto the lunar surface, and his famous words, "That's one small step for man, one giant leap for mankind," were heard around the world. This Moon landing is considered one of the greatest achievements in the history of space exploration.

Week Nine

However, not all space missions went as planned. In 1970, **Apollo 13** was on its way to the Moon when an oxygen tank exploded, putting the astronauts in serious danger. The crew was about 200,000 miles from Earth on the third day of the mission when the accident occurred. The mission had to be aborted, but thanks to the teamwork of the astronauts and **NASA** engineers on Earth, all three astronauts were able to return safely. The story of Apollo 13 is a powerful example of problem-solving, teamwork, and courage in space.

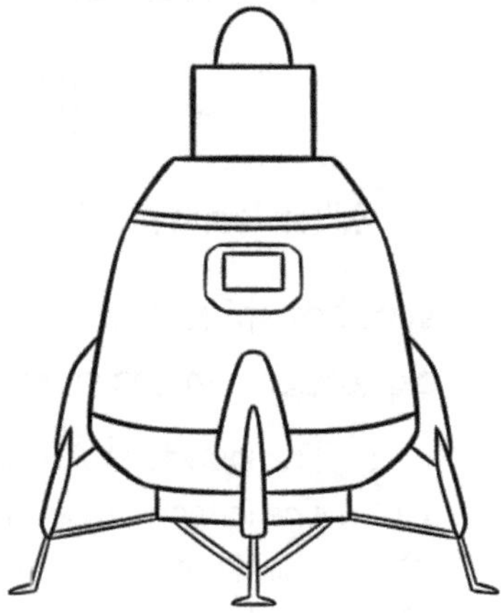

Space exploration did not stop with the Moon. In 1990, NASA launched the **Hubble Space Telescope,** a powerful telescope that orbits Earth and takes detailed pictures of the universe. Hubble has helped scientists see distant galaxies that are trillions of miles away, discover new planets, and learn more about stars and other objects in space.

In summary, space exploration has been an incredible journey. As humans have used telescopes to study the universe and launched satellites into space, we have learned more about the mysteries beyond Earth. New missions are being planned all the time, and the future of space exploration holds many exciting discoveries yet to come.

Name: ____________________

THE HISTORY OF SPACE EXPLORATION

After learning about the history of space exploration, answer the following questions.

1. **What event marked the beginning of space exploration in the passage?**
   a) The launch of Apollo 11
   b) The launch of Sputnik 1
   c) The discovery of the Van Allen Belts
   d) The launch of the Hubble Space Telescope

2. **Who was Laika?**
   a) The first human to orbit Earth
   b) An American astronaut
   c) A scientist who worked on Sputnik
   d) The first dog sent into orbit around Earth

3. **What was the significance of Explorer 1?**
   a) It discovered the Van Allen radiation belts
   b) It discovered galaxies very far away
   c) It carried the first human to the Moon
   d) It was the first spacecraft to orbit the Moon

4. **Which mission successfully landed the first humans on the Moon?**
   a) Apollo 13
   b) Sputnik 1
   c) Apollo 11
   d) Explorer 1

5. **Who was the first person to set foot on the Moon?**
   a) Yuri Gagarin
   b) Alan Shepard
   c) Neil Armstrong
   d) Sally Ride

6. **What happened during the Apollo 13 mission?**
   a) An oxygen tank exploded, but the astronauts returned safely
   b) The mission successfully landed on the Moon
   c) The mission discovered new planets
   d) It discovered the Van Allen Belts

7. **What is the Hubble Space Telescope used for?**
   a) Launching astronauts into space
   b) Taking pictures of the universe
   c) Landing on other planets
   d) Communicating with satellites

8. **What is the main idea of the passage?**
   a) The history of NASA
   b) The dangers of space exploration
   c) Important milestones in the history of space exploration
   d) The discovery of new planets and galaxies

Week Nine

Day 2

Name: ______________________

PLAN YOUR OWN SPACE MISSION

Imagine you are the mission commander planning your own space exploration mission. Use your creativity and knowledge to design a successful mission.

MISSION OBJECTIVE

What is the goal of your space mission? Choose one of the following or come up with your own idea and write it down.

- ☐ Landing on Mars
- ☐ Studying a distant moon
- ☐ Exploring an asteroid
- ☐ Searching for life on one of Jupiter's moons
- ☐ Launching a new space telescope
- ☐ Your own idea: ______________________

____

____

DESCRIBE THE MISSION OBJECTIVE:

Example: My mission will explore the surface of Mars to reach for signs of past or present life and collect soil samples.

____

____

____

____

____

____

____

____

____

EQUIPMENT AND SUPPLIES

What equipment or supplies will you bring? Think about what you need to accomplish your mission.

- TOOLS (Examples: Drills, cameras):

____

____

- LIFE SUPPORT (Example: oxygen tanks):

____

- FOOD AND WATER:

____

- OTHER SUPPLIES:

____

____

____

Name: ______________________________

PLAN YOUR OWN SPACE MISSION

CHALLENGES AND SOLUTIONS

What challenges might your mission face? How will you overcome them?

CHALLENGE 1: Running out of Oxygen

SOLUTION: ______________________________

____

CHALLENGE 1: Equipment Failure

SOLUTION: ______________________________

____

CHALLENGE 3: ______________________________

____

SOLUTION: ______________________________

____

____

CREW SELECTION

Who will be on your team? Consider the skills needed for your mission.

ASTRONAUT 1:

____

ASTRONAUT 2:

____

ASTRONAUT 3:

____

Why did you choose these crew members?

____

____

____

____

MISSION SUCCESS

How will you know if your mission is successful? What are the signs of success? One example of success is that all crew members returned safely to Earth.

____

____

____

____

____

____

NAME YOUR SPACESHIP:

____

DRAW AND COLOR YOUR SPACESHIP

Name: ____________________

VOCAB MATCH

Match each word with its definition. Write the correct letter in the space provided.

1. Satellite ________
2. Orbit ________
3. Astronaut ________
4. Rocket ________
5. Launch ________
6. Spacecraft ________
7. Telescope ________
8. Mission ________
9. Explore ________
10. Discovery ________

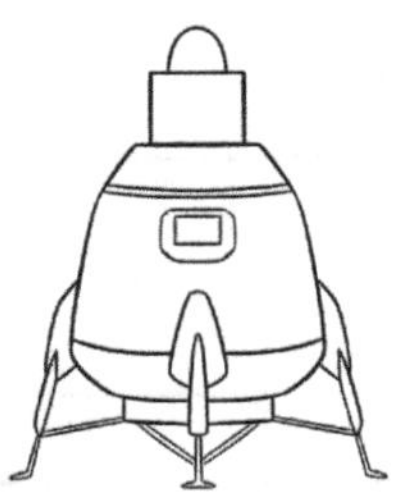

A — A person who is trained to travel and work in space

B — To travel in or study an unfamiliar area to learn more about it

C — The path an object takes as it moves around a planet, star, or moon

D — A vehicle or device designed to travel or operate in outer space

E — An object that orbits around a planet or another celestial body

F — An instrument used to observe distant objects by collecting light

G — A vehicle designed to propel itself by ejecting exhaust gas from one end

H — The act of finding or learning something for the first time, often in a scientific context

I — A specific task or goal assigned to a spacecraft, such as landing on the Moon or exploring another planet

J — The process of sending a spacecraft, satellite, or rocket into space

Name: ______________________

SPACE EXPLORATION TIMELINE

Use the key events provided to create a timeline of significant moments in space exploration history. Then, write a brief description or fact about each event.

Key Events

- Explorer 1 launched
- Sputnik 1 launched
- Apollo 13 mission
- Hubble Space Telescope launched
- Sputnik 2 launched
- Apollo 11 Moon landing

OCTOBER 4, 1957
Event: ______________________
Fact: ______________________

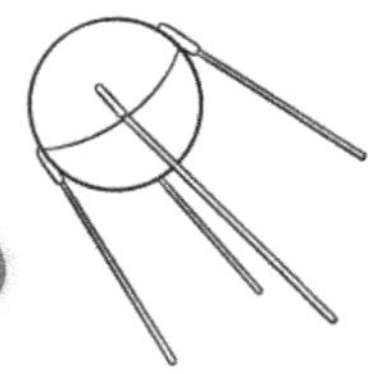

NOVEMBER 3, 1957
Event: ______________________
Fact: ______________________

JANUARY 31, 1958
Event: ______________________
Fact: ______________________

JULY 20, 1969
Event: ______________________
Fact: ______________________

APRIL 11-17, 1970
Event: ______________________
Fact: ______________________

APRIL 24, 1990
Event: ______________________
Fact: ______________________

Week Nine

Day 5

REFLECTION QUESTIONS

1. Which space exploration event do you think was the most significant? Explain your reasoning using what you've learned.

2. What do you think the future of space exploration might include?

With each new **FACT LEARNED**, you get closer to **UNCOVERING** the mysteries of the

UNIVERSE.

Answer keys

Physical Science - Answer Keys

Week One

States of Matter

Page 10 – Day 1

1. Solid
2. Liquid
3. Liquid
4. Gas
5. Solid
6. Solid
7. Liquid
8. Gas

Page 11 – Day 2

1. b)
2. D)
3. A
4. B
5. Atoms
6. The container it is in
7. Answers may vary. Examples may include freezing water, melting ice, boiling water, or water vapor cooling

Page 12 – Day 3

1. C
2. D
3. F
4. A
5. E
6. B
7. Gas
8. Solid
9. Liquid

Page 13 – Day 4

1. C
2. P
3. P
4. C
5. P
6. C
7. C
8. C
9. P
10. P

Page 14 – Day 5

Sample answer: This happens because heat can change the state of matter. When chocolate or candle wax gets hot, it melts and turns from a solid into a liquid. When it cools down, it becomes a solid again because it loses heat. This is a physical change because the substance stays the same. Other substances that do the same thing are butter, ice, and crayons.

Week Two

Mixtures and Solutions

Page 17 – Day 1

1. Mixture
2. Mixture
3. Solution
4. Solution
5. Mixture
6. Solution
7. Solution
8. Mixture

Page 18 – Day 2

1. B
2. A
3. A
4. B
5. You can't see the salt in ocean water because the salt has dissolved.
6. Answers may vary. These are some examples: Soda, dish soap, hot chocolate, lemonade
7. The student will write the definition of a mixture in their own words. When you combine two or more substances and they do not completely mix, it is called a mixture.

Page 19 – Day 3

1. F
2. C
3. E
4. G
5. A
6. B
7. D

Page 20 – Day 4

1. Solution
2. Mixture
3. Solution
4. Solution
5. Mixture
6. Mixture

Page 21 – Day 5

Sample answer. Student response may vary.

Lemonade is made using water, sugar, and lemon juice. The water is the solvent because it dissolves the other ingredients. The sugar is a solute because it dissolves into the water. If more sugar is added than the lemonade can dissolve, the extra sugar will sink to the bottom and not dissolve. This happens because the solution has reached its solubility limit.

Physical Science - Answer Keys

Week Three

Force and Motion

Page 25 – Day 1

1. C
2. A
3. B
4. C
5. Sample answer: Force can change the motion of an object by making it start moving, stop moving, or change direction. For example, when you kick a soccer ball, the force from your foot causes the ball to move.
6. Gravity is a force that pulls objects toward Earth. It causes objects to fall to the ground and keeps people and objects from floating away.

Page 27 – Day 2

1. 3rd
2. 1st
3. 2nd
4. 3rd
5. 2nd
6. 1st

Page 29 – Day 3

1. Pull
2. Friction
3. Pull
4. Gravity
5. Push
6. Gravity
7. Friction
8. Friction

Page 30 – Day 4

1. B
2. H
3. E
4. A
5. F
6. J
7. D
8. I
9. C
10. G

Page 31 – Day 5

Sample Response: The type of surface affects how much friction an object experiences. Rough surfaces create more friction, while smooth surfaces create less friction. When there is more friction, it is harder for an object to move. For example, it is harder to push a box across a carpet because the carpet is rough and creates more friction. It is easier to push the same box across a tile floor because the surface is smooth and has less friction. Another example is riding a bike. Riding on grass is harder than riding on pavement because grass creates more friction.

Week Four

Simple Machines

Page 36 - Day 1

1. Pulley
2. Lever and wheel and axle
3. Inclined plane
4. Inclined plane
5. Wheel and axle
6. Wedge
7. Wheel and axle
8. Screw
9. Pulley and lever
10. Screw and inclined plane
11. Wedge and lever
12. Lever

Page 37 - Day 2

1. A
2. C
3. B
4. D
5. Answers may vary. Examples: light bulb, jar lid, drill
6. Inclined plane
7. Wedge

Page 38 - Day 3

1. H
2. C
3. E
4. K
5. B
6. F
7. L
8. D
9. A
10. J
11. G
12. I

Page 39 - Day 4

Answers will vary depending on what students find. Examples may include:

Inclined Plane: ramp, stairs, slide
Lever: seesaw, crowbar, hammer pulling out a nail
Screw: light bulb socket, screw, jar lid
Pulley: flagpole, window blinds with a cord, elevator system
Wedge: knife, doorstop, axe
Wheel and Axle: bicycle wheels, steering wheel, cart

Physical Science - Answer Keys

Page 40 - 41 - Day 5

Answers and inventions will vary. Focus on understanding, not artistic skills. Creativity encouraged.

Week Five

Electricity

Page 45 - Day 1

1. A series circuit has only one path for electricity to flow. All the parts are connected in a line. If one bulb goes out, the electricity cannot flow anymore, so all the bulbs turn off. A parallel circuit has more than one path for electricity to flow. Each bulb is on its own path. If one bulb goes out, the other bulbs can still stay on because electricity can travel through the other paths.

2. Series
3. Parallel
4. Parallel
5. Series

Page 46 - Day 2

1. B
2. D
3. B
4. A
5. A resistor is something in a circuit that uses electrical energy to do work. A light bulb is an example of a resistor.
6. One disadvantage of a series circuit is that if one bulb goes out, the entire circuit stops working.
7. No, the kitchen lights will not be affected because homes use parallel circuits. Each light has its own path, so electricity can still flow even if one bulb burns out.

Page 47 - Day 3

1. C
2. C
3. I
4. I
5. C
6. I
7. C
8. I

Page 48 - Day 4

1. G
2. B
3. I
4. E
5. H
6. C
7. A
8. D
9. F

Page 49 - 50 - Day 5

Sample student response: I invented a night light for my hallway. It turns on when someone walks by so people can see in the dark.
My invention uses a parallel circuit. This means the light will still work even if one bulb goes out. Electricity has more than one path to follow.
I used conductors like wires to let electricity flow. I used insulators like plastic to keep people safe. This invention helps people see better at night.

Week Six

Light and Thermal Energy

Page 53 - Day 1

1. B
2. D
3. C
4. A
5. Reflection
6. Absorption
7. Refraction
8. Refraction
9. Reflection
10. Absorption

Page 55 - Day 2

1. Translucent
2. Transparent
3. Translucent
4. Opaque
5. Opaque
6. Translucent
7. Transparent
8. Opaque
9. Transparent

Page 57 – Day 3

1. A
2. C
3. B
4. A
5. When you heat up a pot of soup, the heated liquid rises up, and the cooler liquid sinks down.
6. Drawings will vary, but expected examples:

- Conduction: touching a hot spoon or stove
- Convection: boiling water or warm air rising
- Radiation: sun, fire, or heat felt without touching

Physical Science - Answer Keys

Page 58 - Day 4

1. C
2. I
3. D
4. K
5. A
6. G
7. F
8. J
9. B
10. E
11. H

Page 59 - 60 - Day 5

Sample student response:
One example of reflection I see in my daily life is a mirror in my bathroom. When I look in the mirror, light hits the mirror and bounces back to my eyes, allowing me to see my reflection. Another example of reflection is sunlight reflecting off a car window. The light bounces off the shiny surface and can make it hard to see.

One example of refraction is when a straw looks bent in a glass of water. The light bends as it moves from the air into the water, which makes the straw appear bent. Another example of refraction is when light passes through eyeglasses. The light bends as it moves through the lenses, helping people see more clearly.

Week Seven

Kinetic and Potential Energy

Page 62 - Day 1

1. Kinetic
2. Potential
3. Kinetic
4. Potential
5. Kinetic
6. Kinetic
7. Potential
8. Kinetic
9. Potential

Page 63 - Day 2

1. A
2. A
3. B
4. C
5. C
6. Sample response: "Tug of War" is a game where two teams pull on opposite ends of a rope. The rope starts with potential energy due to tension but no kinetic energy. As teams pull, potential energy converts to kinetic energy, increasing as the teams exert more force. The game ends when one team pulls the other across a line, showing the balance between kinetic and potential energy.

Page 65 - Day 3

1. C (lowest point, fastest speed)
2. A (highest point)
3. C
4. A
5. B
6. A

Page 66 - Day 4

1. E
2. I
3. A
4. D
5. G
6. B
7. H
8. J
9. C
10. F

Page 67 - 68 - Day 5

Answers will vary depending on their roller coaster drawing.

Week Eight

Sound Waves

Page 71 - Day 1

1. B
2. C
3. A
4. C
5. Sound waves are created when an object vibrates. The vibrations make the air around the object move in waves. These waves travel through the air to our ears and make our eardrums vibrate. Our brain then understands the vibrations as sound.
6. Amplitude is how tall the sound wave is and it affects how loud the sound is. A bigger amplitude means a louder sound. Frequency is how often the sound waves happen, and it affects the pitch. A higher frequency makes a higher-pitched sound, and a lower frequency makes a lower-pitched sound.

Physical Science - Answer Keys

Page 72 – Day 2

1. B
2. A
3. B
4. B
5. B (the higher the frequency, the higher the pitch)
6. 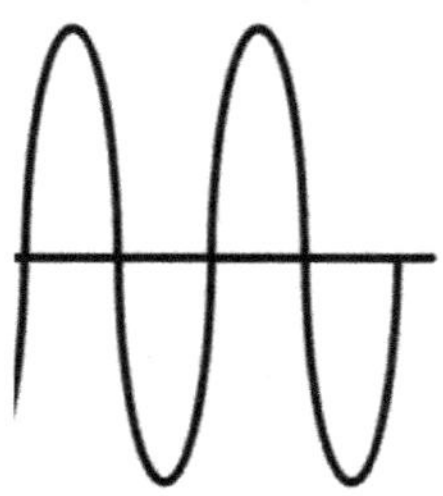

Page 73 – Day 3

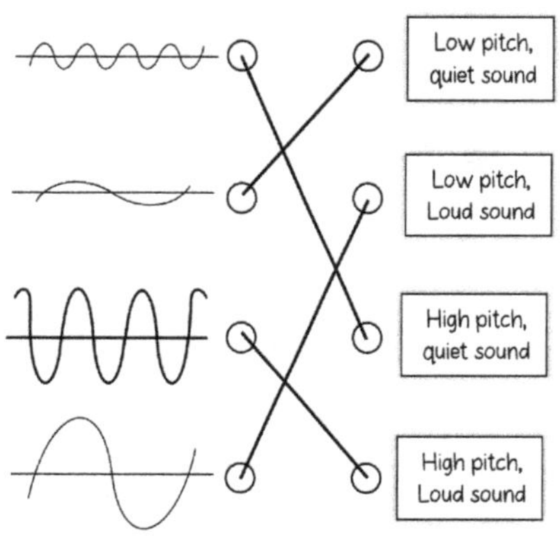

Page 74 – Day 4

1. E
2. B
3. H
4. D
5. J
6. A
7. G
8. F
9. I
10. C

Page 75 – Day 5

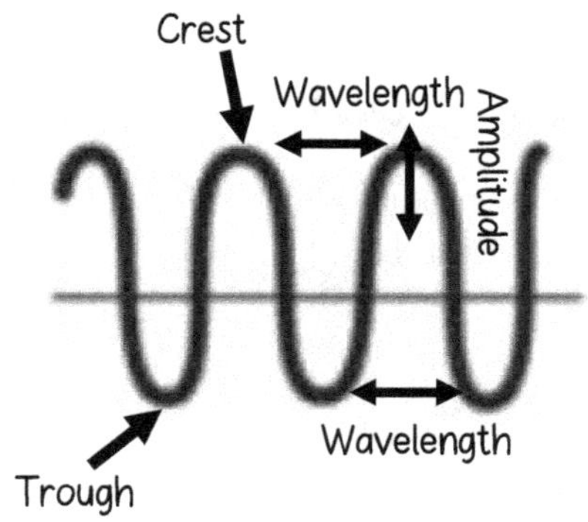

Week Nine

Magnetism

Page 79 – Day 1

1. C
2. B
3. D
4. A
5. Sample student response: Magnets play a crucial role in many everyday objects. According to the passage, magnets are used in compasses, fans, cars, and computers to function properly. For example, a compass uses Earth's magnetic field to point north, fans often have electric motors that contain magnets, cars use magnets in various components like sensors and motors, and computers rely on magnets in their hard drives and other parts.

Page 80 – Day 2

1. Thumbs-down (non-magnetic)
2. Thumbs-up (magnetic)
3. Thumbs-up (magnetic)
4. Thumbs-down (non-magnetic)
5. Thumbs-up (magnetic)
6. Thumbs-down (non-magnetic)
7. Thumbs-down (non-magnetic)
8. Thumbs-up (magnetic)
9. Thumbs-down (non-magnetic)
10. Thumbs-up (magnetic)

Page 81 – Day 3

1. Attract
2. Repel
3. Attract
4. Attract
5. Repel
6. Attract
7. Attract
8. Repel
9. Attract

Page 82 – Day 4

1. C
2. F
3. A
4. H
5. B
6. G
7. D
8. I
9. E

Page 83 – Day 5

Example response: If magnets stopped working, many everyday items would stop working too. Compasses would not point north, so it would be hard to find directions. Fans, cars, and computers would not work properly because they use magnets. Speakers and headphones would not make sound. Life would be harder because many machines and devices we use every day depend on magnets.

Life Science - Answer Keys

Week One

Food Chains and Food Webs

Page 86 – Day 1

1. Herbivore
2. Decomposer
3. Omnivore
4. Carnivore
5. Carnivore
6. Producer
7. Herbivore
8. Decomposer

Page 87 – Day 2

1. B
2. A
3. C
4. B

Page 88 – Day 3

1. D
2. B
3. C
4. A
5. E
6. Students may choose their own animals for each category. Check that the animals listed match the correct category. Also check that the illustrations show the foods those animals eat. For example, if a student lists a bunny and an elephant as herbivores, they may draw carrots or grass to represent their food.

Page 89 – Day 4

1. Phytoplankton and seaweed
2. Blue whale and killer whale because they are not shown being eaten by any other organisms in the food web.
3. 8

Page 90 – Day 5

Responses may vary. Look for an understanding that changes to one organism can affect many others in a food chain or food web.

Sample student response: A disruption in one part of a food chain or food web can affect other organisms because all living things depend on each other for food and energy. If one organism disappears, the animals that eat it may not have enough food. At the same time, the organisms it used to eat might grow too much. This can cause the whole ecosystem to become unbalanced.

Week Two

Biomes

Page 92– Day 1

1. Tundra
2. Rainforest
3. Marine
4. Desert
5. Freshwater
6. Grassland

Page 93 – Day 2

1. D
2. B
3. A
4. C
5. Describe it- (Sample response): Freshwater biomes include rivers, lakes, and ponds. The water is not salty. Plants like reeds and water lilies grow there, and animals such as fish, frogs, turtles, and insects live in freshwater habitats.

Page 94 – Day 3

1. B
2. E
3. A
4. D
5. C
6. Students research tropical and temperate rainforests. Below is an example of a filled-in chart. (Answers will vary.)

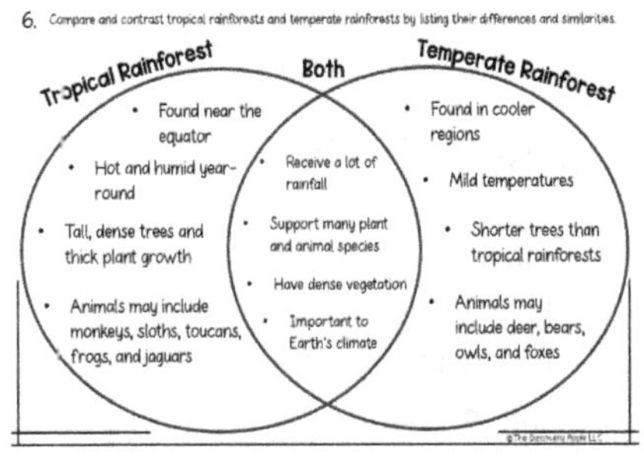

View up-close example – page 331

Page 95 – Day 4

Reflection question example student response: (Any answer showing location awareness is correct.)

I learned that tropical rainforests are located near the equator.

I learned that deserts are not only hot but are found in specific areas of the world.

I learned that tundra is found near the North Pole.

Page 96 – Day 5

Answers will vary. Sample Student response: I would investigate how desert plants store water and protect themselves from heat. I would also study how animals find food and water. I want to learn how adaptations like burrowing or being active at night help them survive.

Life Science - Answer Keys

Week Three

Animal Adaptations

Page 98 - Day 1

1. Instinctive
2. Instinctive
3. Learned
4. Instinctive
5. Learned
6. Learned

Page 99 – Day 2

1. Tusks
2. Trunk
3. To protect them from the sun and insects
4. They are trying to cool off because they are very hot.

Page 100 – Day 3

1. Physical
2. Behavioral
3. Physical
4. Behavioral
5. Behavior
6. Physical

Page 101 – Day 4

1. H
2. C
3. G
4. E
5. A
6. F
7. J
8. B
9. I
10. D

Page 102 – Day 5

Student Sample Response:
I chose a polar bear. One physical adaptation a polar bear has is thick fur and sharp claws. The thick fur keeps the polar bear warm in cold Arctic temperatures, and the sharp claws help it catch and hold onto slippery prey like seals.

One behavioral adaptation a polar bear uses is waiting quietly near holes in the ice where seals come up to breathe. This behavior helps the polar bear save energy and surprise its prey. Both the physical and behavioral adaptations help the polar bear find food and survive in its cold environment.

Week Four

Animal Classification

Page 104 - Day 1

1. Reptile
2. Amphibian
3. Fish
4. Mammal
5. Bird
6. Insect

Page 105 – Day 2

Students' answers may vary. Any correct examples with correct classifications should be accepted.

Warm-blooded animals:
Dog – Mammal
Penguin– Bird
Giraffe– Mammal
Cat– Mammal
Eagle – Bird
Cold-blooded animals:
Snake – Reptile
Frog – Amphibian
Stingray – Fish
Lizard – Reptile
Ladybug – Insect

Page 106 – Day 3

1. Vertebrate
2. Vertebrate
3. Invertebrate
4. Vertebrate
5. Invertebrate
6. Invertebrate

Page 107 – Day 4

1. F
2. I
3. D
4. A
5. G
6. B
7. J
8. E
9. H
10. C

Page 108 - 109 – Day 5

Sample student response:
I discovered a new animal called the jungle glider. It has thick fur, strong legs, and long claws. The thick fur helps keep the animal warm, and the claws help it grab branches and catch insects. One physical adaptation the jungle glider has is its sharp claws, which help it hold onto trees and find food.
The jungle glider also has a behavioral adaptation. It climbs trees at night to search for insects and fruit. This behavior helps it avoid predators and find food more easily.
The jungle glider is a mammal because it has fur and gives birth to live young. It is warm-blooded because it can keep its body temperature the same in different weather. It is also a vertebrate because it has a backbone. These adaptations help the jungle glider survive in its jungle environment.

Life Science - Answer Keys

Week Five

Plant and Animal Cells

Page 113 – Day 1

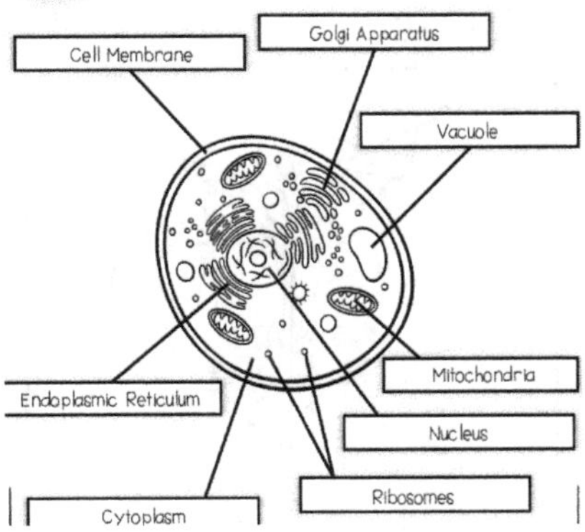

View up-close example – page 331

Page 114 – Day 2

1. D
2. B
3. C
4. A
5. Endoplasmic Reticulum
6. Golgi Apparatus
7. Ribosomes
8. Cells need proteins because proteins help build structures, repair damage, and carry out chemical reactions that keep the cell functioning.

Page 115 – Day 3

Cell Membrane – The layer that controls what comes in and out of the cell.
Vacuole – Holds water and other materials
Chloroplast – Uses energy from the sun to make food for the plant cell.
Golgi Apparatus – Sorts and processes proteins
Ribosomes – Makes the proteins
Endoplasmic Reticulum – Takes molecules where they need to go
Nucleus – Controls the activities of the cell
Cytoplasm – Holds and protects the cell's organelles
Cell Wall – The layer that gives structure and support in plant cells

Page 116 – Day 4

1. Both
2. Plant Cell
3. Plant Cell
4. Plant Cell
5. Both
6. Animal Cell

Page 117 – Day 5

Sample student response:
(Look for mention of sunlight, making food, or energy)
I am a plant cell, and chloroplasts are very important to me. Chloroplasts help me use sunlight to make my own food through photosynthesis. This food gives me energy to grow and stay healthy. Without chloroplasts, I would not be able to make my own food. I would not have enough energy to survive or grow properly. Chloroplasts help me stay alive and support the whole plant.

Week Six

Human Body Systems

Page 121 - Day 1

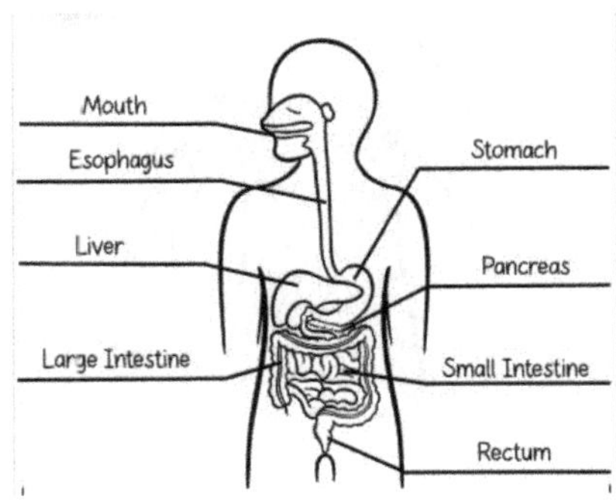

1. The digestive system breaks down food into nutrients the body can use.
2. Most nutrient absorption happens in the small intestine

Page 122 – Day 2

1. B
2. A
3. C
4. A

Page 123 – Day 2

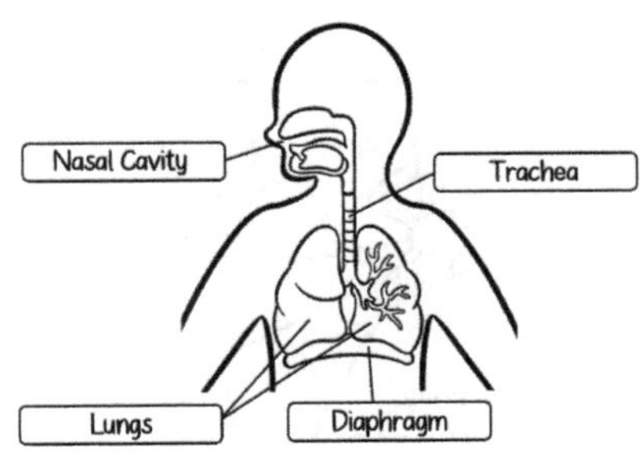

1. b
2. d
3. a

Life Science - Answer Keys

Page 125 – Day 3

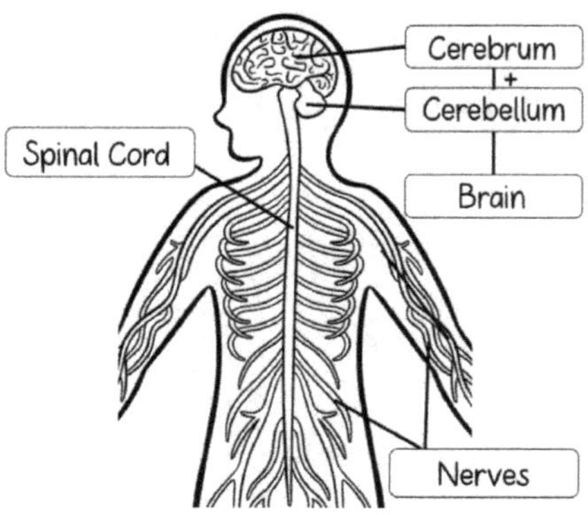

1. B
2. A
3. C

Page 126 – Day 3

1. H
2. D
3. G
4. E
5. A
6. F
7. C
8. I
9. B

Page 127 – Day 4

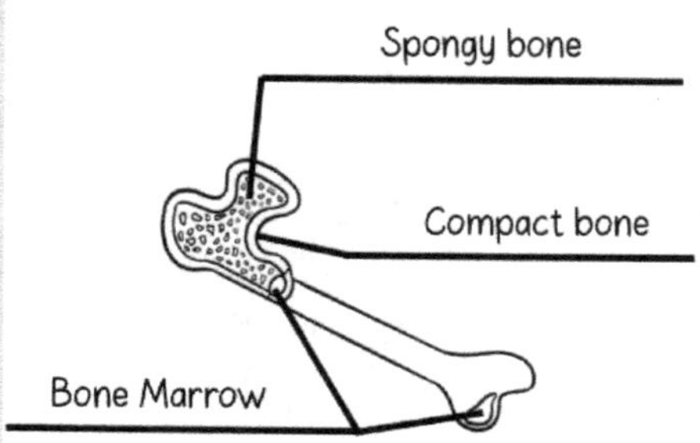

Page 128 – Day 4

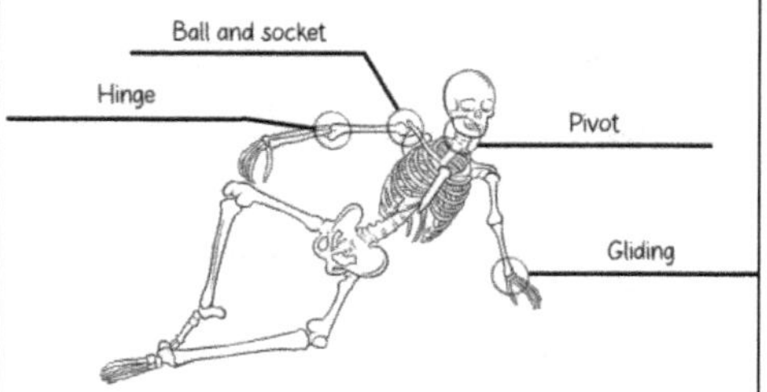

View up-close example – page 332

1. C
2. A
3. B

Page 129 – Day 5

1. B
2. B
3. C

Page 130 – Day 5

Example Student response:
When I run during recess, my heart beats faster to pump blood through my body. The blood carries oxygen to my muscles so they can keep moving. The circulatory system helps my body get the oxygen it needs so I don't get tired too quickly.

Teacher Tip: Students do not need to use all vocabulary words to earn full credit. Look for understanding of how the heart and blood help the body during movement.

Page 131 – Extra Practice

1. Skeletal System
2. Nervous System
3. Respiratory System
4. Muscular System
5. Digestive System
6. Circulatory System

Week Seven

Parts of a Plant

Page 134 – Day 1

Draw It:

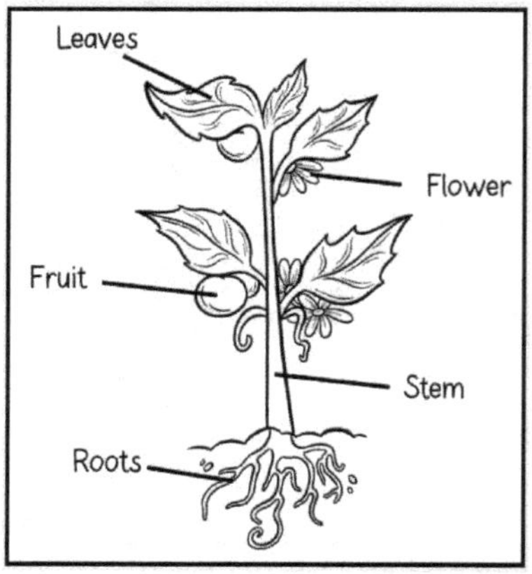

1. B
2. D

Page 135 – Day 2

1. D
2. A
3. B
4. C

Page 136 – Day 3

1. D
2. I
3. A
4. G
5. C
6. H
7. B
8. F
9. E

Life Science - Answer Keys

Page 137 – Day 4

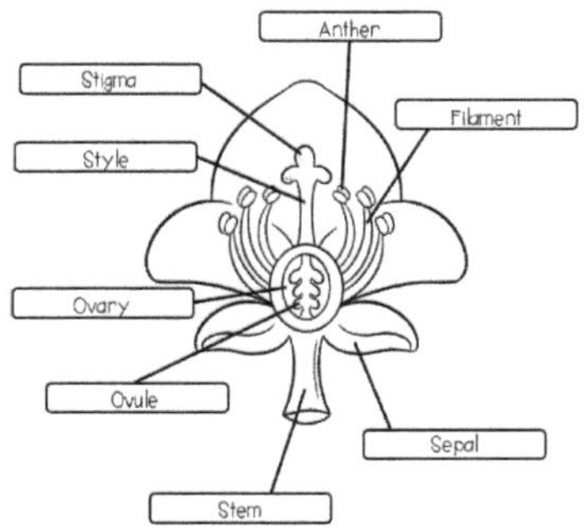

View up-close example – page 332

Challenge Answer:
Student will circle the anther and draw a star around the stigma.

Page 138 – Day 5

Sample student answer:
Plants have thorns or spines to help protect themselves. Thorns can stop animals from eating the plant. This helps the plant survive and grow. Spines can also help keep the plant safe from damage by other animals.

Week Eight

Plant Life Cycle

Page 140 – Day 1

Page 141 – Day 2

The student will describe each term in their own words and draw a picture to show their explanations. Example responses:

Seed – A seed is the part of a plant that can grow into a new plant.
Germination – when a seed begins to wake up and sprout
Seedling – a baby plant that has just started to grow
Young plant – bigger than a seedling but still growing and getting ready to become a full-grown plant
Mature plant – fully-grown plant that has reached its maximum size producing flowers, fruits, and seeds
Pollination – how plants make "babies" by transferring pollen from one flower to another

Page 142 – Day 3

1. F
2. B
3. E
4. I
5. G
6. A
7. C
8. H
9. D

Page 143 – Day 4

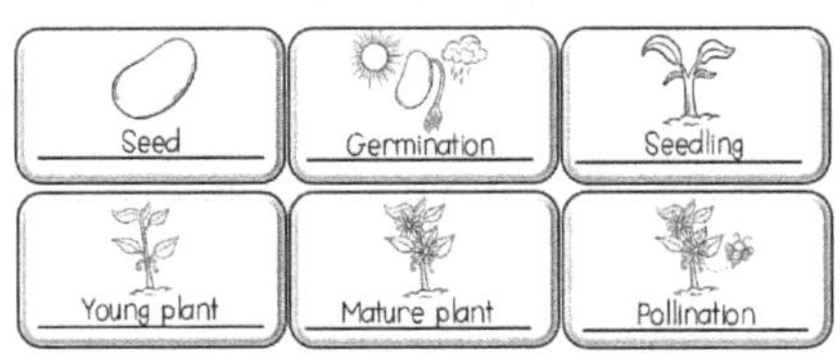

View up-close example – page 333

1. B
2. C
3. C
4. D

Page 144 – Day 5

Sample student answer:
Pollination is important in the plant life cycle because it helps plants make new seeds. When pollination happens, pollen moves from the male part of a flower to the female part. This allows the flower to make seeds that can grow into new plants. Pollination can happen with the help of bees, butterflies, or even the wind. Without pollination, plants would not be able to reproduce and survive.

Week Nine

Photosynthesis

Page 146 – Day 1

1. False
2. Photosynthesis
3. They take the energy from the sun and use it to convert carbon dioxide and water into food
4. Glucose (or sugar)
5. Sunlight, water, and carbon dioxide
6. Plants breathe out oxygen and we need oxygen to breathe.
7. Chlorophyll
8. Through xylem from the soil

Page 147 – Day 2

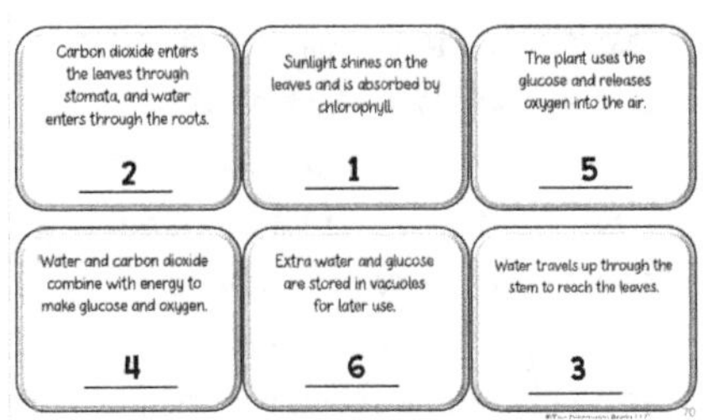

View up-close example – page 333

Life Science - Answer Keys

Page 148 – Day 3

1. E
2. C
3. F
4. A
5. D
6. G
7. B

Page 149 – Day 4

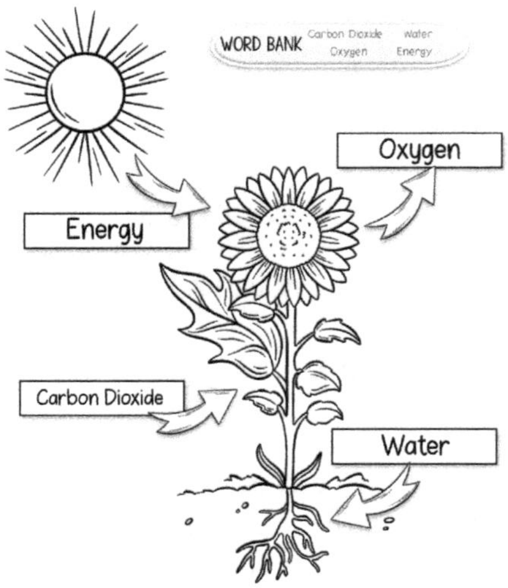

Page 150 – Day 5

Sample student response:

If plants could not make their own food through photosynthesis, many living things would be affected. Plants would not be able to grow or survive because they need photosynthesis to make food. Without plants, there would be less oxygen in the air, which people and animals need to breathe. Animals that eat plants would not have food, and animals that eat other animals would also be affected. This would harm the food chain and make it hard for living things to survive.

Page 94 – Day 3

6. Compare and contrast tropical rainforests and temperate rainforests by listing their differences and similarities.

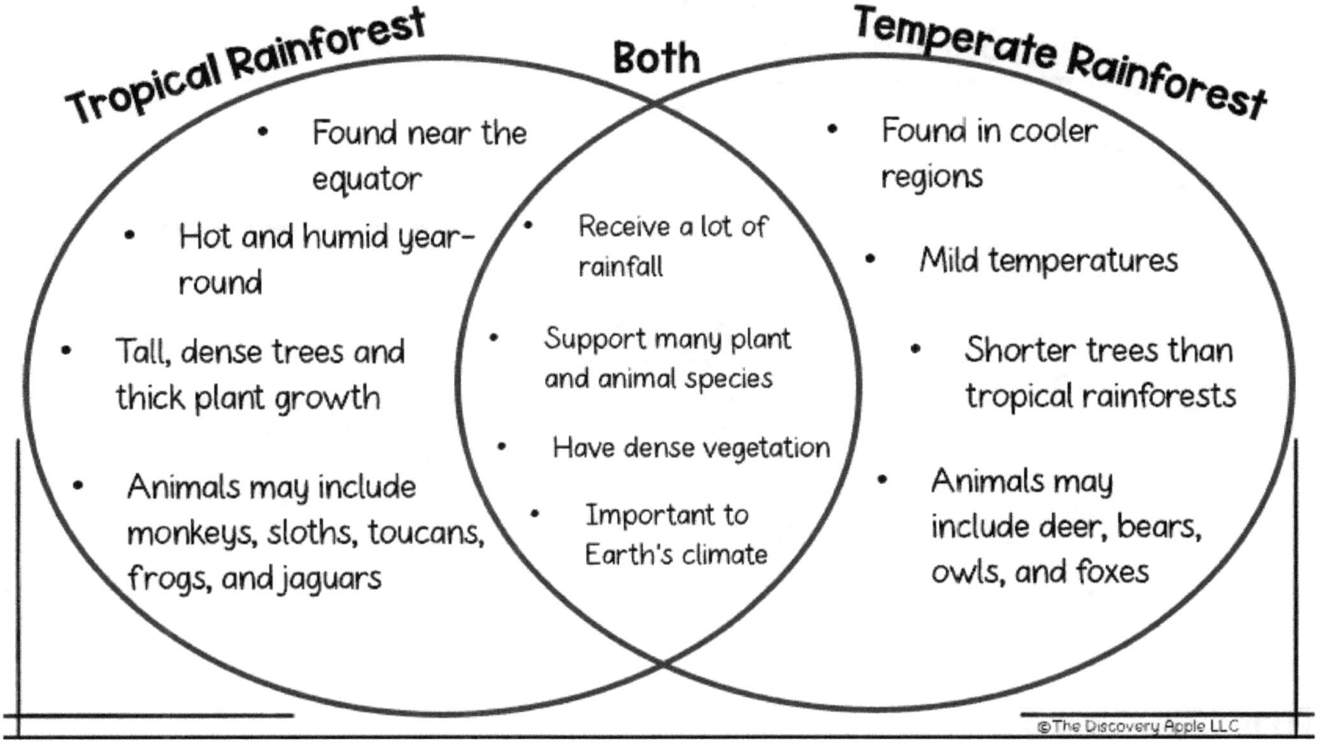

Page 113 – Day 1

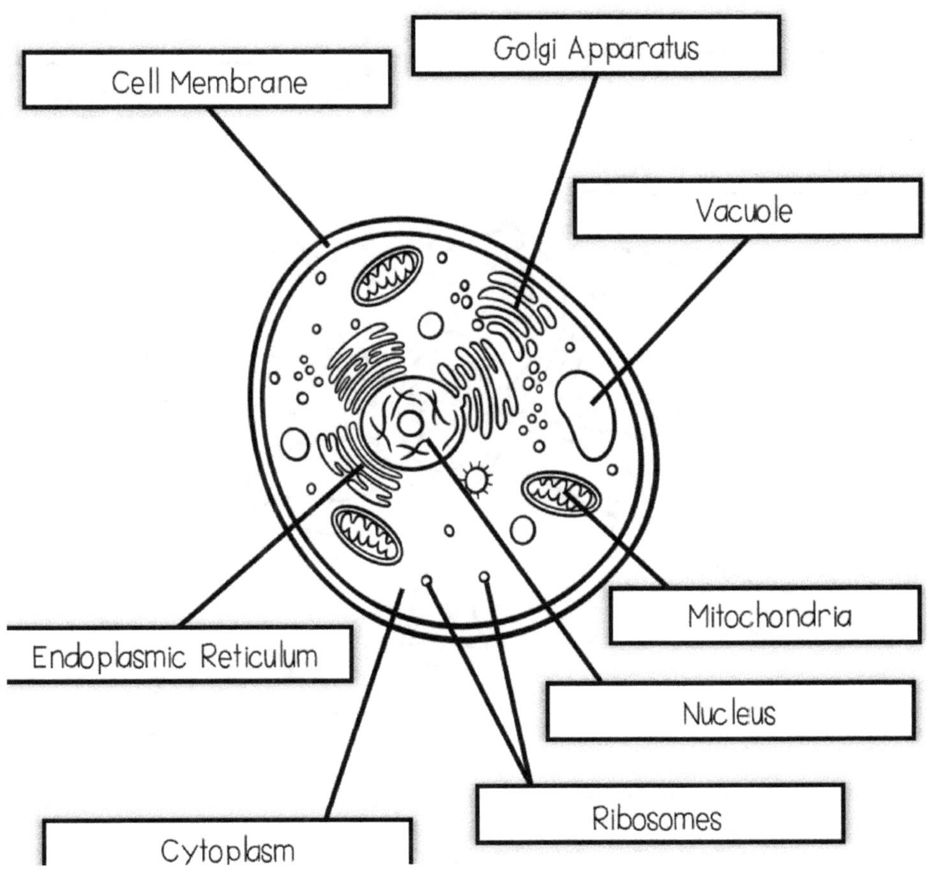

Page 128 – Day 4

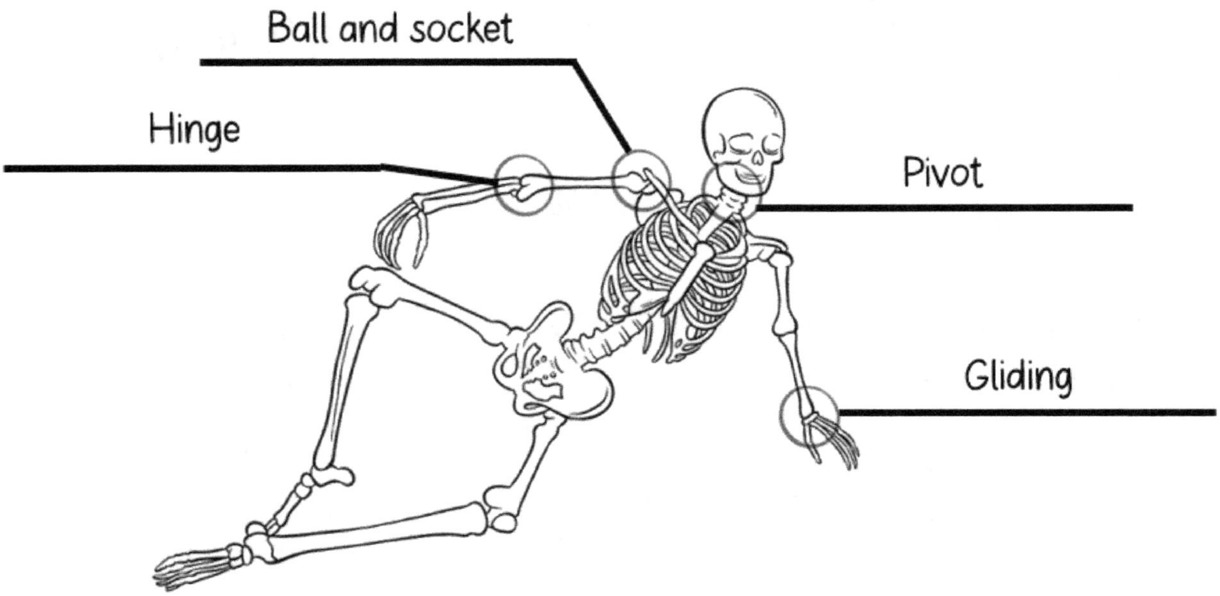

Page 137 – Day 4

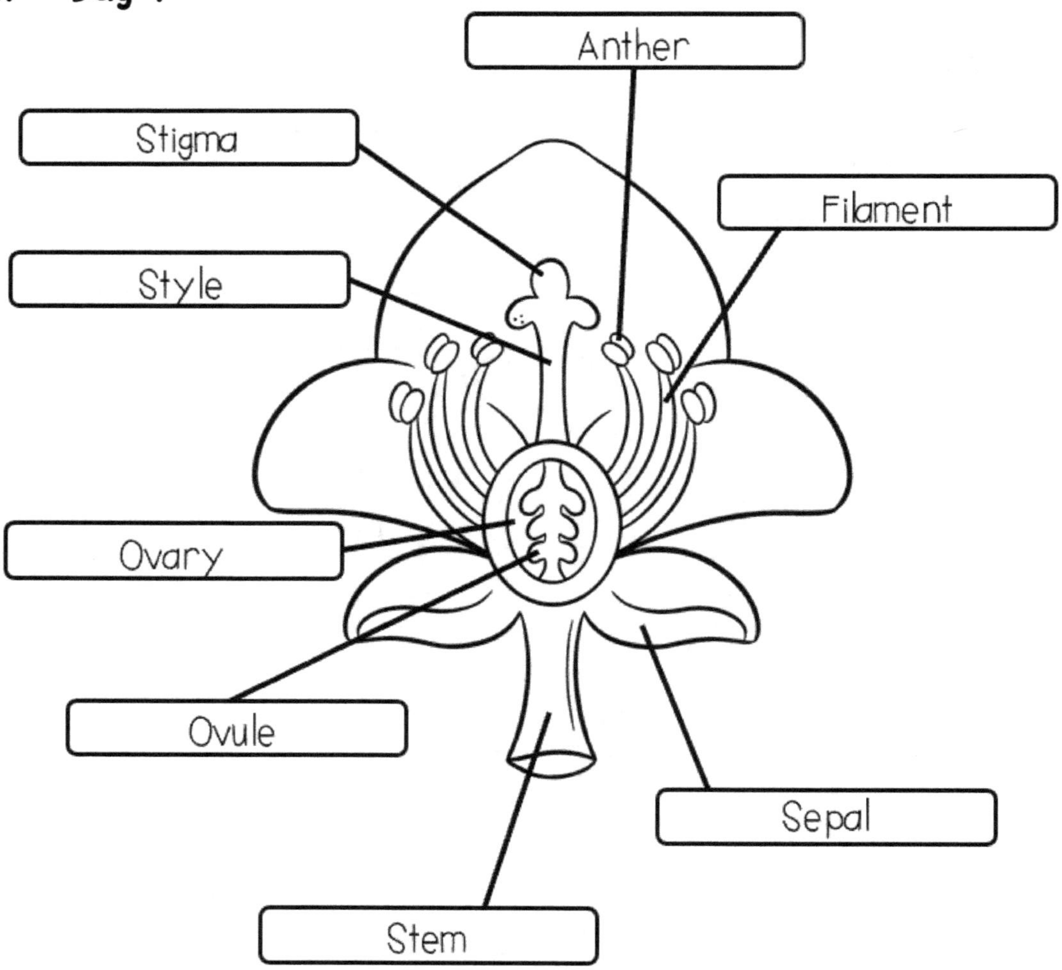

Page 143 – Day 4

Seed

Germination

Seedling

Young plant

Mature plant

Pollination

Page 147 – Day 2

Carbon dioxide enters the leaves of a plant through their tiny holes known as stomata. Water enters the plant through its roots.

2

Sunlight shines on the leaves of a plant and is absorbed by a green pigment called chlorophyll.

1

The plant uses the glucose but releases the oxygen into the air.

5

The water and carbon dioxide combine with the stored energy to make food (glucose) and oxygen.

4

The extra water and glucose are stored in the vacuoles of the plant cells for use later on when needed.

6

Water travels all the way through the stem to reach the leaves.

3

Earth Science - Answer Keys

Week One

Rocks and Minerals

Page 156 - Day 1

1. B
2. B
3. A
4. C
5. Sediments are small pieces of rocks, minerals, and organic materials that settle at the bottom of bodies of water or gather on land.
6. In earth's crust
7. Limestone being exposed to high heat and pressure
8. Minerals

Page 157 – Day 2

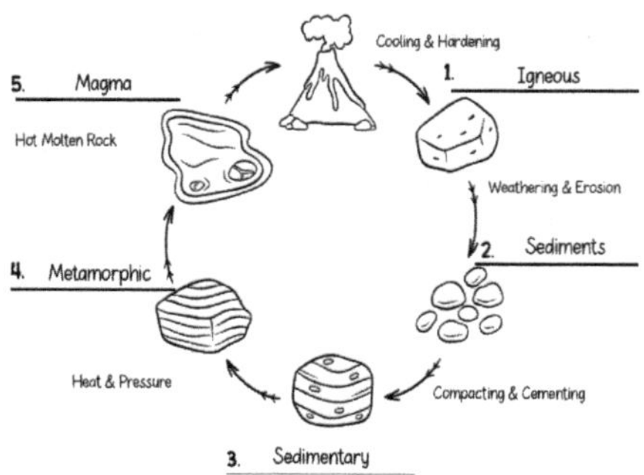

View page 340 for a larger image.

1. Igneous
2. Sediments
3. Sedimentary
4. Metamorphic
5. Magma

Page 159 – Day 3

1. Erosion
2. Deposition
3. Weathering
4. Weathering
5. Deposition
6. Erosion
7. Deposition
8. Weathering
9. Erosion

Page 160 – Day 4

1. D
2. J
3. G
4. B
5. F
6. A
7. H
8. E
9. C
10. I

Page 161 – Day 5

Sample response:
If I were a geologist studying a new rock formation, I would examine the rock's characteristics. If the rock has crystals, it may be igneous. If it has layers, it may be sedimentary. If it has bands or is very hard, it may be metamorphic. I would also think about how the rock may have changed over time. For example, if I find a metamorphic rock, it means it was once another type of rock that changed because of heat and pressure. I would look for evidence of processes such as melting, cooling, compacting, and heating. These clues help explain how the rock formed and transformed. Fossils found in a rock can also show that it was once sedimentary and formed from layers of sediment.

Week Two

Soil

Page 165 – Day 1

1. D
2. D
3. B
4. C
5. It takes a long time for soil to form because rocks slowly break down through weathering, and layers of soil build up over many years.
6. If I had a garden, I would prefer loamy soil because it has a balanced mix of sand, silt, and clay. This balance allows it to hold water and nutrients well while still draining properly, providing the ideal environment for plants to grow healthy and strong.

Page 166 – Day 2

1. Silt
2. Loam
3. Sand
4. Clay
5. Clay
6. Sand
7. Loam
8. Silt
9. Clay

Page 168 – Day 3

1. Humus
2. Topsoil
3. Subsoil
4. Parent Material
5. Bedrock
6. Bedrock
7. Subsoil
8. Humus

Earth Science - Answer Keys

Page 169 – Day 4

1. G
2. A
3. F
4. D
5. B
6. C
7. E
8. J
9. H
10. I

Page 170 – Day 5

Sample response:
It is important for farmers and gardeners to take care of the soil because healthy soil helps plants grow strong and produce food. Soil provides nutrients and water that plants need to survive. Farmers and gardeners can keep soil healthy by adding compost, planting cover crops, watering properly, and not removing too many nutrients from the soil.

Week Three

Fossils

Page 174 – Day 1

1. C
2. A
3. B
4. B
5. Example Response: Fossilization happens when an organism is buried by sediment and, over time, minerals replace the original material, turning it into rock.
6. Example response: Body fossils are the actual remains of plants or animals, like bones or teeth. Trace fossils are marks left behind, such as footprints or burrows.

Page 175 – Day 2

1. Mold
2. True-form
3. Trace
4. True-form
5. Cast
6. Mold
7. Cast
8. Trace
9. Mold
10. Trace
11. Cast
12. True-form

Page 176 – Day 3

3 — The sediment surrounding the skeleton thickens and begins to turn to stone.

6 — The fossil is exposed on Earth's surface because of wind and rain.

5 — Minerals enter the mold and fills the cavity, which forms a cast.

2 — The skeleton is buried by sediment.

1 — An animal dies and its soft parts rot away, leaving only the skeleton.

4 — The skeleton dissolves, leaving a mold behind.

Page 177 – Day 4

1. C
2. G
3. E
4. A
5. J
6. F
7. I
8. D
9. H
10. B

Page 178 – Day 4

Sample response:
It is important to study different types of fossils because each type gives us different information about Earth's past. Body fossils show what plants and animals looked like, such as their bones or shells. Trace fossils show how organisms lived, moved, or behaved, like footprints or burrows. Mold and cast fossils show the shape of organisms that no longer exist. When scientists study all types of fossils together, they can better understand what lived on Earth long ago and how environments have changed over time.

Week Four

Landforms & Bodies of Water

Page 184 – Day 1

1. Island
2. Sand dune
3. Peninsula
4. Valley
5. Mountain
6. Plateau
7. Canyon
8. Plain
9. Hill
10. Delta
11. Mesa
12. Archipelago

Earth Science - Answer Keys

Page 185 - Day 2

1. River
2. Lake
3. Canal
4. Sea
5. Pond
6. Ocean
7. Waterfall
8. Gulf
9. Glacier

Page 186 – Day 3

1. Canyon
2. Hills
3. Island
4. Peninsula
5. Plateau
6. Valley
7. Gulf
8. Waterfall
9. River
10. Lake
11. Canal
12. Ocean

Page 187 – Day 4

1. B
2. D
3. A
4. C
5. Example response: A waterfall forms when a river flows over hard rock and softer rock underneath erodes faster. This creates a steep drop where water falls from a height. One famous waterfall is Niagara Falls.
6. Example Response: A plateau and a mesa are similar because they both have flat tops and steep sides. They are different because a plateau is much larger, while a mesa is smaller.

Page 188 – Day 5

Sample response:
Living near a lake or pond has both benefits and challenges. One benefit is having access to fresh water for drinking, fishing, and activities like swimming and boating. Lakes and ponds also provide habitats for wildlife, making them great places to observe animals and plants. However, there are challenges as well. Flooding can happen during heavy rains and may affect nearby homes and crops. Mosquitoes and other insects are also more common near water, which can be annoying for people who live nearby.

Week Five

The Water Cycle

Page 191 – Day 1

1. A
2. B
3. B
4. C
5. Example response: The water cycle is important because it continuously moves and recycles water around Earth. It provides fresh water for plants, animals, and people to survive. Without the water cycle, life on Earth would not have enough water.
6. Example response: Clouds form during condensation when water vapor cools and turns into liquid droplets. Clouds hold water until they become heavy enough to release precipitation, such as rain or snow.

Page 192 – Day 2

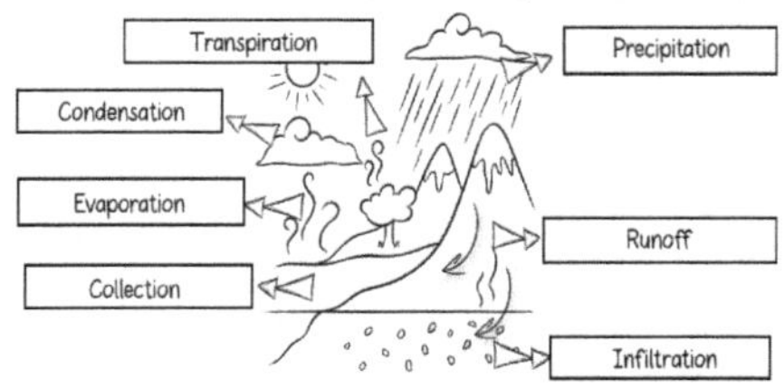

View page 340 for a larger image.

1. Sun
2. Gas
3. Rocks ; infiltration
4. Precipitation ; evaporate

Page 193 – Day 3

1. <u>Transpiration</u> : Evaporation of water from plant leaves
2. <u>Condensation</u>: gas turns into liquid
3. <u>Precipitation</u>: Snow falling to the ground
4. <u>Evaporation</u>: liquid turning into gas
5. Pre-labeled
6. <u>Runoff</u>: Precipitation that did not get absorbed into the soil travels to the ocean
7. <u>Infiltration</u>: rain water soaks into the ground through soil and rocks
8. Student will draw the water cycle.

Page 194 – Day 4

1. F
2. D
3. A
4. B
5. J
6. E
7. I
8. G
9. H
10. C

Earth Science - Answer Keys

Page 195 – Day 5

Sample Response:
The water cycle impacts ecosystems and the environment by providing water that plants, animals, and people need to survive. Precipitation supplies fresh water to rivers, lakes, and soil, helping plants grow and animals find drinking water. Evaporation and condensation help move water around Earth, which affects weather and climate. The water cycle also helps refill groundwater and keeps ecosystems balanced by constantly recycling water.

Week Six

Weather & Climate

Page 199 – Day 1

☺	TORNADOES	HURRICANES
How do they form?	Over land	Over warm water in tropical oceans
How big are they?	No more than a few hundred feet wide	Up to several hundred miles wide
How long do they last?	About an hour at the most	For days or even weeks
How strong are the winds?	As fast as 300 miles per hour	Less than 180 miles per hour
How common are they?	800 – 1000 each year	10 each year

View page 341 for a larger image.

6. 6. Yes, people usually have enough time to plan and evacuate for a hurricane because hurricanes can be predicted several days in advance. Weather scientists can track hurricanes over the ocean, which gives people time to prepare, gather supplies, and leave the area if needed.
7. People usually have very little time to prepare for a tornado, often only about 15 to 30 minutes. Tornadoes can form quickly and without much warning.
8. Once a hurricane hits land, it begins to weaken. Hurricanes lose strength because they no longer have warm ocean water to keep them strong.

Page 201 – Day 2

1. D
2. F
3. B
4. E
5. A
6. C
7. Barometer
8. Rain gauge
9. Wind vane
10. Anemometer
11. Hygrometer
12. Thermometer

Page 203 – Day 3

1. B
2. C
3. B
4. B
5. D
6. A
7. C
8. D

Page 206 – Day 4

View page 341 for a larger image.
Make sure the student colored in the fronts and pressure systems the correct colors (not shown in the image).

Page 207 – Day 5

1. Example response: Weather patterns affect what activities I can do during the year. In the summer, hot weather makes it easier to swim, play outside, and go on vacations. In the winter, cold weather or snow can stop outdoor sports and make people stay inside more. Rainy weather can cancel games or outdoor plans, while sunny days make it easier to be active. Weather also affects daily routines, like wearing warm clothes in winter or waking up earlier to avoid heat in the summer.

2. A sudden change in climate could greatly affect the environment and people living there. If an area became much hotter or drier, plants might not get enough water and could die. Animals might have to move to new places to survive. If an area became colder or wetter, flooding or freezing temperatures could damage homes and crops. People would need to change how they live, what they grow, and how they prepare for weather.

Earth Science - Answer Keys

Week Seven

Volcanoes

Page 212 – Day 1

1. B
2. C
3. A
4. C
5. Magma is melted rock that is found inside the Earth. Lava is magma that has reached the Earth's surface during an eruption.
6. A shield volcano is wide with gentle slopes and is formed by flowing lava. A composite volcano is taller, steeper, and made from layers of lava, ash, and rock.
7. A dormant volcano has not erupted in a long time, but it may erupt again in the future. It is sometimes called a sleeping volcano.

Page 213 – Day 2

1. D
2. I
3. G
4. A
5. J
6. F
7. C
8. H
9. B
10. E

Page 214 – Day 3

Responses will vary.

Page 215 – Day 4

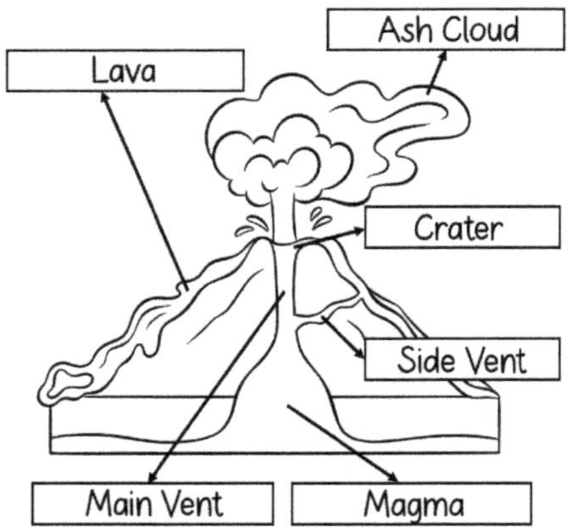

Page 216 – Day 5

<u>Sample response</u>:
Volcanic eruptions can be very dangerous to people and the environment. One danger is lava, which is extremely hot and can destroy homes, roads, and plants. Another danger is volcanic ash, which can make it hard to breathe and can damage crops and animals. Ash can also block sunlight and cover large areas of land. Volcanic eruptions can release harmful gases that are dangerous for people and animals to breathe. Explosions from eruptions can throw rocks into the air and cause injuries. Eruptions can also change the land by creating new mountains or covering areas with ash and lava. While volcanoes can help create new land over time, eruptions can be very destructive in the moment.

Page 217 – The Volcano Experiment

Sample Student Response:

<u>Why do you think this eruption happened?</u>
The eruption happened because the baking soda and vinegar reacted together and made a lot of gas. The gas built up and pushed the mixture out of the bottle like a volcano erupting.

<u>If you add more vinegar, how will it change the eruption? Explain.</u>
If I add more vinegar, the eruption will last longer because there will be more liquid to react with the baking soda. The eruption might not be taller, but it could keep bubbling for a longer time.

<u>If you add more baking soda, how will it change the eruption? Explain.</u>
If I add more baking soda, the eruption might be bigger and foam up more. There would be more gas made, so the eruption could be stronger and overflow faster.

Week Eight

Plate Tectonics

Page 222 – Day 1

1. C
2. B
3. A
4. B
5. Scientists study tectonic plate movements to understand and help predict earthquakes, volcanic eruptions, and mountain formation.
6. Sample student answers (answers may vary): They can cause earthquakes that damage buildings and roads.
7. B
8. A

Earth Science - Answer Keys

Page 223 – Day 2

	DESCRIPTION	ILLUSTRATION	EXAMPLES
TRANSFORM	A fault zone where two plates slide past each other	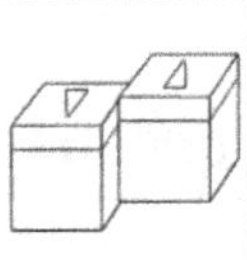	• San Andreas Fault • New Zealand's Alpine Fault • North America's Queen Charlotte Fault
DIVERGENT	Two plates move away from one another	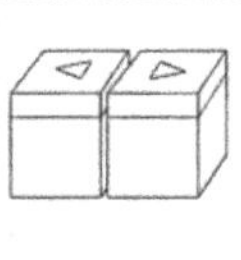	• Mid Atlantic Ridge • Red Sea Rift • Great Rift Valley • East African Rift • West Antarctic Rift
CONVERGENT	Two plates push together. When one plate moves under the other, it is called subduction.	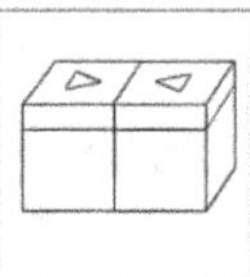	• Collision between Eurasian and Indian Plate forming the Himalayas • Subduction of Nazca plate beneath South American Plate to form the Andes

Page 224 – Day 3

1. Pacific Plate
2. North American Plate
3. Nazca Plate
4. South American plate
5. Antarctic Plate
6. African Plate
7. Eurasian Plate
8. Australian Plate

Page 225 – Day 4

1. E
2. H
3. A
4. F
5. G
6. D
7. B
8. I
9. C

Page 226 – Day 5

1. Even though tectonic plates move very slowly, they are very large and heavy. Over time, energy builds up where the plates meet. When the plates suddenly shift or break free, that stored energy is released, causing earthquakes. Volcanoes happen when plate movement allows magma to rise to the surface.

2. If tectonic plates stopped moving, there would be fewer earthquakes and volcanoes. New mountains would stop forming, and Earth's surface would change very slowly. Over time, this could affect landforms, habitats, and the way living things adapt to their environment.

Week Nine

Natural Resources

Page 230 – Day 1

1. B
2. D
3. A
4. B
5. Sample student answer: Renewable resources can be replaced naturally, like sunlight and trees. Non-renewable resources cannot be replaced quickly and can run out, like oil and coal.
6. Sample student answer: It is important to manage natural resources so they do not run out and so future generations can use them too.
7. B (False – People do not make natural resources.)
8. A

Page 231 – Day 2

1.

View page 342 for a larger image.

2. Coal is a non-renewable resource because it takes millions of years to form and cannot be replaced quickly once it is used.

Page 232 – Day 3

1. Renewable
2. Renewable
3. Non-renewable
4. Renewable
5. Non-renewable
6. Non-renewable
7. Renewable
8. Renewable
9. Non-renewable

Page 233 – Day 4

1.	G	5.	C
2.	B	6.	F
3.	E	7.	A
4.	H	8.	D

Page 234 – Day 5

Human activities can affect natural resources by using too much water, cutting down too many trees, and burning fossil fuels. When people overuse resources, they can run out or become polluted. We can protect natural resources by conserving water, recycling, using renewable energy, and planting new trees. Taking care of natural resources helps make sure they are available for the future.

The Rock Cycle

Page 157 – Day 2

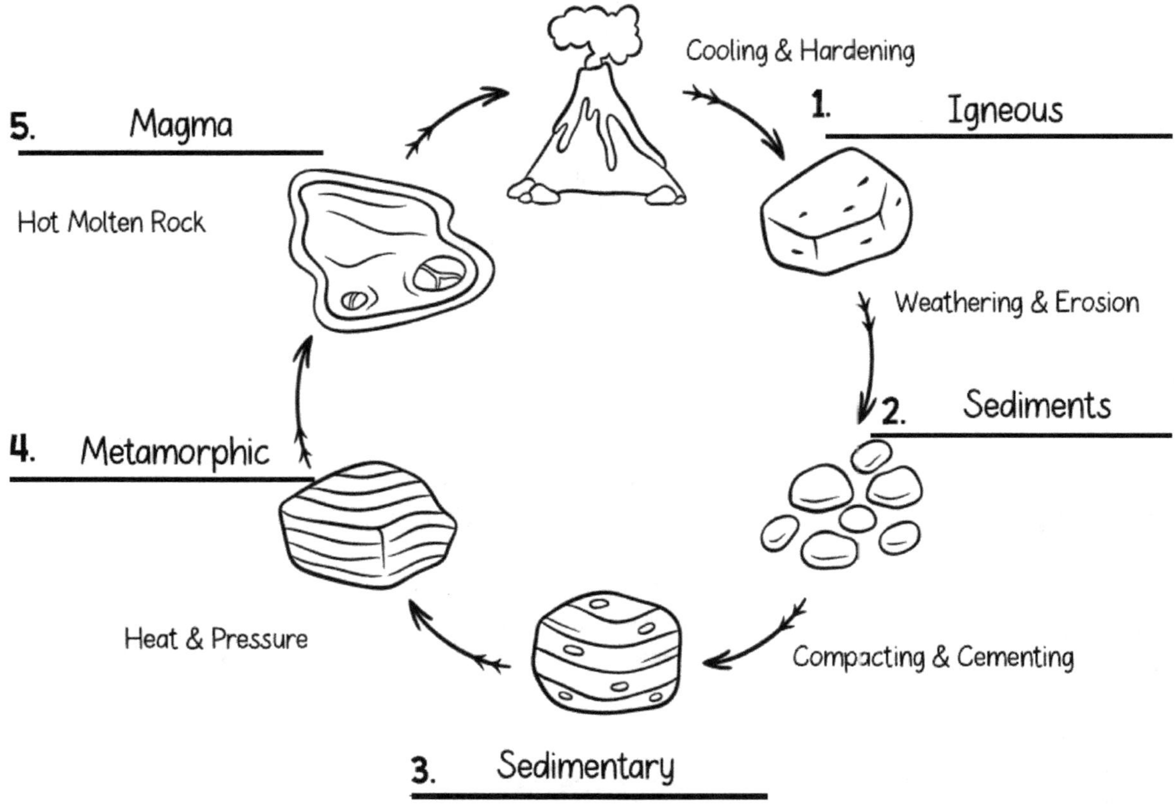

Page 192 – Day 2

The Water Cycle

Tornadoes vs. Hurricanes

Page 199 – Day 1

	TORNADOES	HURRICANES
How do they form?	Over land	Over warm water in tropical oceans
How big are they?	No more than a few hundred feet wide	Up to several hundred miles wide
How long do they last?	About an hour at the most	For days or even weeks
How strong are the winds?	As fast as 300 miles per hour	Less than 180 miles per hour
How common are they?	800 – 1000 each year	10 each year

Page 206 – Day 4

Weather Map

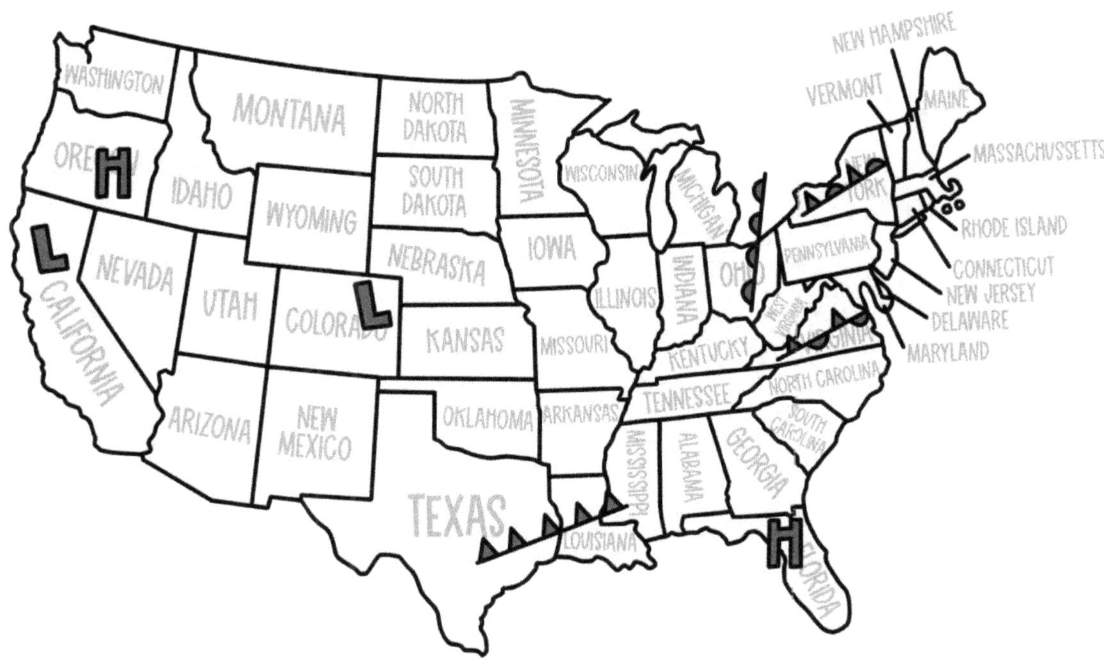

Coal Formation

Page 231 – Day 2

STEP 1	STEP 2	STEP 3
	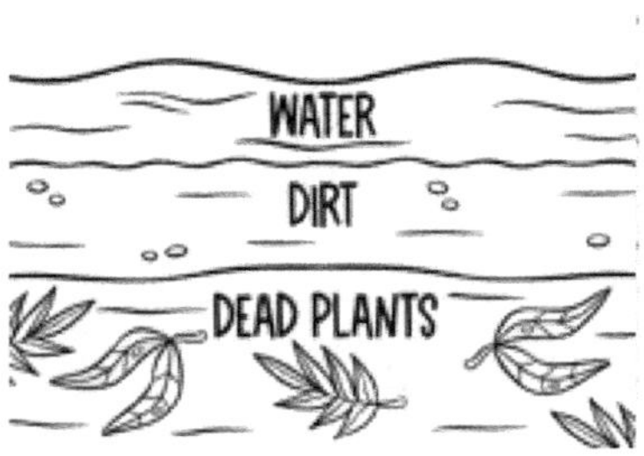	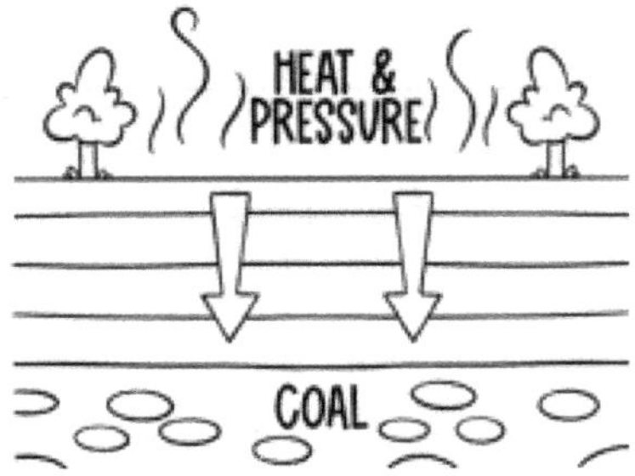
Millions of years ago, many large plants died in swamps.	Over time, the dead plants were buried under water and dirt.	Heat and pressure continued to build up as more layers formed, transforming the dead plants into coal

Space Science - Answer Keys

Week One

The Planets

Page 240 - Day 1

1. Uranus
2. Mercury
3. Earth
4. Saturn
5. Jupiter
6. Mars
7. Neptune
8. Venus

Page 241 – Day 2

1. Mercury – terrestrial
2. Venus – terrestrial
3. Earth – terrestrial
4. Mars - terrestrial
5. Jupiter – Gas Giant
6. Saturn - Gas Giant
7. Uranus - Gas Giant
8. Neptune - Gas Giant

Page 242 – Day 3

1. C
2. B
3. B
4. A
5. False – Mars is farther from the Sun than Earth
6. Terrestrial planets are smaller, rocky, and have solid surfaces. Gas giants are much larger and are made mostly of gases like hydrogen and helium.
7. Answers will vary. Sample answer: Jupiter is unique because it is the largest planet in the solar system and has a giant storm called the Great Red Spot.

Page 243 – Day 4

1. F
2. D
3. B
4. J
5. E
6. H
7. G
8. A
9. C
10. I

Page 244 – Day 5

Sample student response: If I were planning a mission to another planet, I would choose Mars. Mars is a rocky planet and has the tallest volcano in the solar system. Scientists also think Mars may have had water in the past. I would want to explore Mars to learn if it could have supported life and to study its surface.

Week Two

Moon Phases

Page 248 – Day 1

1. B
2. C
3. B
4. A
5. False – A waxing Moon appears to be getting bigger.
6. Full Moon
7. Because the side of the Moon facing Earth is not lit by the Sun

Page 249 – Day 2

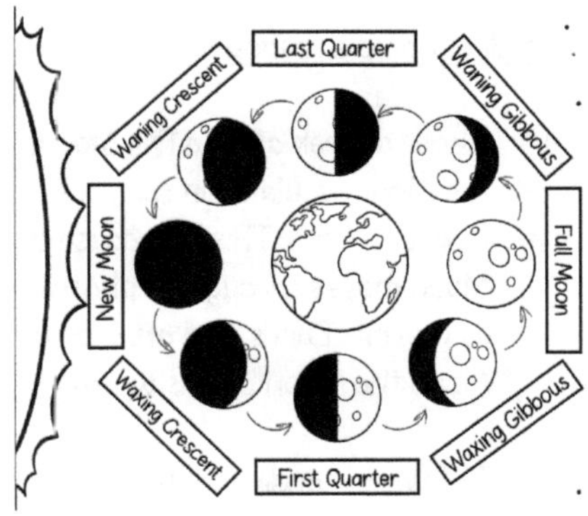

View page 349 for a larger image.

Page 250 – Day 3

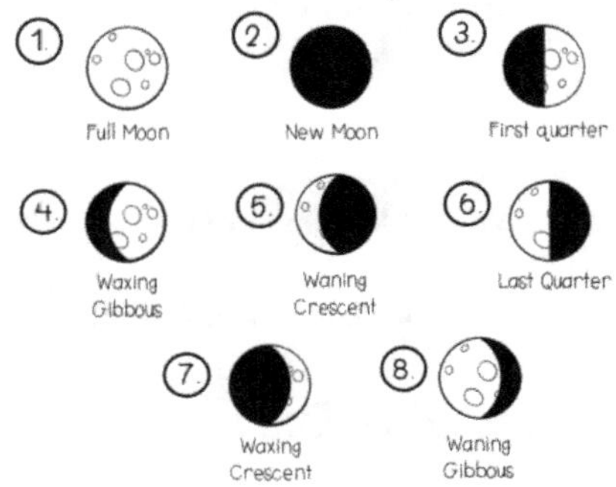

9. B) Waning
10. A) Waxing

Page 251 – Day 4

1. H
2. C
3. I
4. A
5. F
6. D
7. E
8. B
9. G

Space Science - Answer Keys

Page 252 - Day 5

Student Sample response:

1. About a week after a Full Moon, the Moon would be in the Last Quarter phase. This is because the Moon moves through its phases as it orbits Earth, and after a Full Moon the Moon begins waning.

2. If I were standing on the Moon looking at Earth, I would see Earth going through phases instead of the Moon. Earth would appear full when we see a New Moon, and Earth would appear dark when we see a Full Moon.

Week Three

Solar & Lunar Eclipses

Page 260 – Day 1

1. B
2. D
3. A
4. C
5. Sample Answer: It is safe to view a lunar eclipse without special glasses because you are looking at the Moon, not the Sun. During a solar eclipse, looking at the Sun can damage your eyes.

6. Sample Answer: Ancient people did not understand eclipses, so they were scared and believed they were signs from the gods or warnings of danger.

Page 261 – Day 2

1. Sun
2. Moon
3. Penumbra
4. Umbra
5. Earth
6. Moon's orbit
7. Sample Student Response: This diagram shows a solar eclipse. The Moon moves between the Sun and the Earth and blocks the Sun's light. The darkest shadow is called the umbra, and the lighter shadow is called the penumbra. When Earth is in these shadows, a solar eclipse occurs.

Page 262 – Day 3

1. Solar
2. Lunar
3. Solar
4. Solar
5. Lunar
6. Lunar
7. Lunar
8. Solar
9. Lunar
10. Solar

Page 263 – Day 4

1. G
2. B
3. E
4. J
5. A
6. H
7. C
8. F
9. D
10. I

Page 264 – Day 5

Sample Student Response:

A total solar eclipse and a total lunar eclipse both happen when the Sun, Earth, and Moon line up. During a total solar eclipse, the Moon blocks the Sun and the sky becomes dark for a short time during the day. During a total lunar eclipse, the Earth blocks the Sun's light from reaching the Moon, causing the Moon to look dark or reddish at night. Both involve shadows, but they look different and happen at different times.

Week Four

Life Cycle of Stars

Page 268 – Day 1

1. D
2. B
3. C
4. A
5. Sample student response: A massive star and an average star go through different life cycles. An average star, like the Sun, becomes a red giant and then forms a planetary nebula. It ends as a white dwarf. A massive star becomes a red supergiant and then explodes in a supernova. After the supernova, it can become a neutron star or a black hole.

Space Science - Answer Keys

Page 269 – Day 2

PROTOSTAR	RED SUPERGIANT	NEUTRON STAR OR BLACK HOLE
2	4	6
SUPERNOVA	**STELLAR NEBULA**	**MAIN SEQUENCE STAR**
5	1	3

Page 270 – Day 3

Sample captions:
<u>Panel 2</u> – Protostar: Gravity pulls the gas a dust together, and I begin to heat up as a protostar.
<u>Panel 4</u> – Red Giant: I'm running out of hydrogen, so my outer layers expand and I become much bigger!
<u>Panel 5</u> – Planetary Nebula: I shed my outer layers, forming a colorful cloud of gas called a planetary nebula.
<u>Panel 6</u> – White Dwarf: My core is left behind as a white dwarf, slowly cooling over time.

Page 271 – Day 4

1. I
2. C
3. A
4. F
5. E
6. J
7. G
8. K
9. D
10. H
11. B

Page 272 – Day 5

<u>Sample Student Response:</u>

Dear Friend, As a star like our Sun gets older, it runs out of hydrogen fuel and begins to change. It expands into a red giant, becoming much larger and brighter. This change can affect the planets around it by increasing temperatures and possibly changing their orbits. Studying the life cycle of stars helps scientists understand how stars form, change, and affect the universe over time.

Week Five

Constellations

Page 275 – Day 1

1. B
2. A
3. D
4. C
5. B
6. True
7. A
8. C

Page 276 – Day 2

Answers will vary.

Page 279 – Day 3

1. Pegasus
2. Ursa Major
3. Andromeda
4. Canis Major
5. Draco
6. Hercules
7. Cassiopeia
8. Perseus
9. Orion
10. Cygnus

Page 280 – Day 4

1. E
2. C
3. A
4. D
5. F
6. B
7. The swan
8. The captive princess
9. The winged horse

Page 281 – Day 5

<u>Sample Student Response:</u>

Different cultures have different names and stories for the same constellations because people imagine shapes in the stars based on their own traditions and beliefs. Long ago, people used the night sky to tell stories, explain events, and teach lessons. This shows that people see the night sky not just as stars, but as a way to share culture, history, and ideas about the world.

Space Science - Answer Keys

Week Six

Comets, Meteors, and Asteroids

Page 286 - Day 1

1. B
2. A
3. D
4. C
5. Sample student response: I have never seen a meteor shower, but I would like to see one. I think it would be exciting to watch many shooting stars light up the sky at once. People find meteor showers fascinating because they happen quickly and remind us that space is active and full of surprises.

Page 287 - Day 2

	COMET	METEOR	ASTEROID
1. known for having a bright, glowing tail that can be seen from Earth	✓		
2. a small, rocky body that orbits the sun, often found in a belt between Mars and Jupiter			✓
3. often called a "shooting star" when it burns up in Earth's atmosphere		✓	
4. If this object reaches Earth's surface without burning up, it is called a meteorite.		✓	
5. this can be as small as a pebble and can also be hundreds of miles wide			✓
6. mostly made of ice, dust, and rocks	✓		
7. can be seen as a streak of light in the night sky when it burns up in the atmosphere		✓	
8. Develops a coma as it nears the Sun	✓		

Page 288 & 289 - Day 3

1. Comets
2. Coma
3. Away
4. Meteorite
5. Asteroid
6. Orbit
7. Dust
8. Meteors
9. Shower
10. History

C	E	C	A	D	U	E	N	E	C	O	M	A	T
A	S	E	P	S	O	P	E	I	T	I	M	S	G
M	H	A	B	D	T	U	M	A	S	I	I	N	B
O	O	W	D	C	M	E	T	E	O	R	S	L	E
U	W	L	M	A	C	S	R	O	N	E	R	G	H
F	E	A	S	R	P	O	Y	O	I	C	A	L	A
L	R	O	Y	S	R	T	M	R	I	I	T	N	V
A	I	S	P	Y	E	O	E	E	T	D	I	S	I
H	O	R	A	M	D	L	T	T	T	O	O	F	O
E	I	W	E	E	A	B	E	E	I	S	N	S	R
Y	A	S	E	R	T	I	O	O	R	O	F	I	T
D	S	B	T	Y	O	V	R	U	Z	Z	N	I	L
U	H	I	B	O	R	N	I	T	I	O	B	Q	N
S	R	A	V	F	R	Y	T	U	V	R	G	Z	L
T	H	O	J	E	R	Y	E	N	O	A	D	F	R
E	H	K	L	I	U	O	F	X	L	O	B	Y	I

View page 349 for a larger image.

Page 290 - Day 4

1. E
2. C
3. A
4. G
5. J
6. B
7. H
8. F
9. I
10. D

Page 291 - Day 5

Example Response:

Dear Friend,

I learned some cool stuff in science class about asteroids, meteors, and comets, and I wanted to share it with you! They're all space rocks, but they're actually pretty different from each other. First, asteroids are big rocks that orbit the sun, kind of like mini planets. Most of them hang out in a place called the Asteroid Belt, which is between Mars and Jupiter. Some are small, like the size of a car, but others are huge, even bigger than a city! Next, there are meteors. A meteor is what you see when a small rock, called a meteoroid, enters Earth's atmosphere and burns up. It's that bright streak of light in the sky that people call a "shooting star," even though it's not a star at all! If a meteor doesn't burn up completely and actually hits the ground, then it's called a meteorite. Finally, there are comets. Comets are like dirty snowballs made of ice, dust, and rock. They travel around the sun in really long orbits. When a comet gets close to the sun, the heat makes the ice turn into gas, creating a glowing tail that can be millions of miles long! That's what makes comets so cool to watch.

Your friend,

Amir

Week Seven

Gravity

Page 295 - Day 1

1. B
2. C
3. A
4. D
5. Gravity is important in the solar system because it keeps the planets in orbit around the sun. The sun's gravity pulls the planets towards it, preventing them from flying off into space.

Space Science - Answer Keys

Page 296 - Day 2

THE MOON 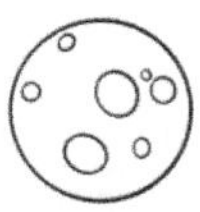 3	A CAR  5
THE SUN  1	AN ASTEROID 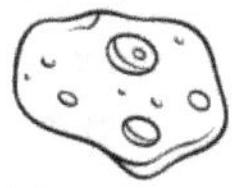 4
A BASEBALL 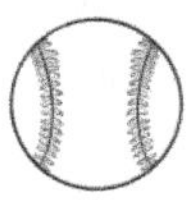 6	EARTH  2

Page 297 - Day 3

1. False
2. True
3. True
4. False
5. True
6. False
7. False
8. True
9. False
10. True

Page 298 - Day 4

1. F
2. B
3. H
4. A
5. C
6. E
7. J
8. D
9. I
10. G

Page 299 - Day 5

1. Sample student response: If Earth's gravity became weaker, people and objects would weigh less. We would be able to jump higher and might have trouble staying balanced when we walk or run. Objects could fall more slowly, and some things might float if gravity was very weak. Moving around would feel harder to control because gravity would not pull us down as strongly.

2. Gravity pulls the ball back down to the ground after we throw it into the air. This helps the ball come back instead of floating away. If there was no gravity, the ball would keep moving through space and never fall back down. It would be very hard to play sports or keep objects near us without gravity.

Week Eight

Earth's Rotation and Revolution

Page 303 - Day 1

1. B
2. A
3. B
4. D
5. Sample student response: From space, I can see that Earth is always moving in two different ways. Earth's rotation is when it spins on its axis, and this is what causes day and night. It takes about 24 hours for Earth to make one full rotation. Earth's revolution is when it travels around the Sun. This movement takes about 365 days and makes one year. Rotation causes day and night, while revolution helps create the seasons.

Page 304 - Day 2

1. True
2. False
3. True
4. True
5. False
6. True
7. False
8. False
9. False
10. True

Page 305 - Day 3

1. Day
2. Year
3. Revolution
4. Night
5. Orbit
6. Sun
7. Rotation
8. Seasons
9. Axis
10. Tilt

Space Science - Answer Keys

Page 306 - Day 4

1. F
2. C
3. H
4. A
5. E
6. G
7. I
8. D
9. B

Page 307 - Day 5

Sample Student Response:

1. If Earth did not rotate on its axis, one side of the planet would always face the Sun and stay in daylight, while the other side would always be dark. This would cause extreme temperatures, with very hot areas on the sunny side and very cold areas on the dark side. We would not have regular day and night, which would make it hard for plants, animals, and people to survive.

2. I would design a planet that rotates very slowly, so each day would last many weeks. This would mean long days and long nights. I would also give the planet a small tilt, so the seasons would not change very much. The weather would stay more the same throughout the year, and people would need to adjust to long periods of sunlight and darkness.

Week Nine

History of Space Exploration

Page 312 - Day 1

1. B
2. D
3. A
4. C
5. C
6. A
7. B
8. C

Page 313 & 314 - Day 2

Answers will vary.

Page 315- Day 3

1. E
2. C
3. A
4. G
5. J
6. D
7. F
8. I
9. B
10. H

Page 316- Day 4

View page 350 for a larger image.

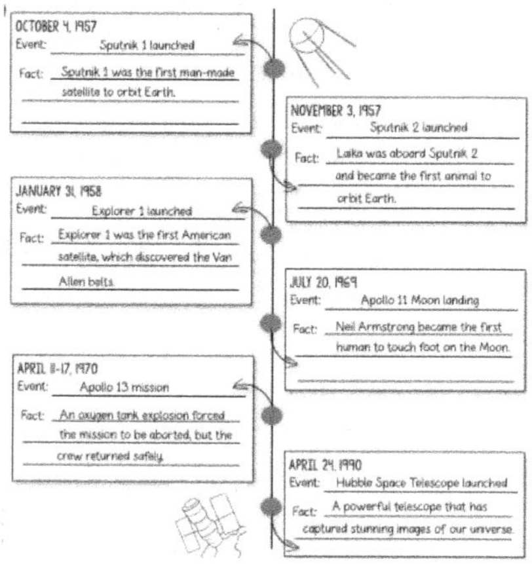

Page 317 - Day 5

Sample student response:

1. I think the Apollo 11 Moon landing was the most significant space exploration event. It was the first time humans landed on the Moon, and Neil Armstrong became the first person to walk on it. This showed that humans could travel far beyond Earth and explore other worlds. It also inspired scientists to continue studying space and planning new missions.

2. I think the future of space exploration might include humans traveling to Mars or building space stations on the Moon. Scientists may also discover new planets and search for signs of life using powerful telescopes. Space exploration will help us learn more about the universe and how to protect Earth.

Page 249 – Day 2

Moon Phases

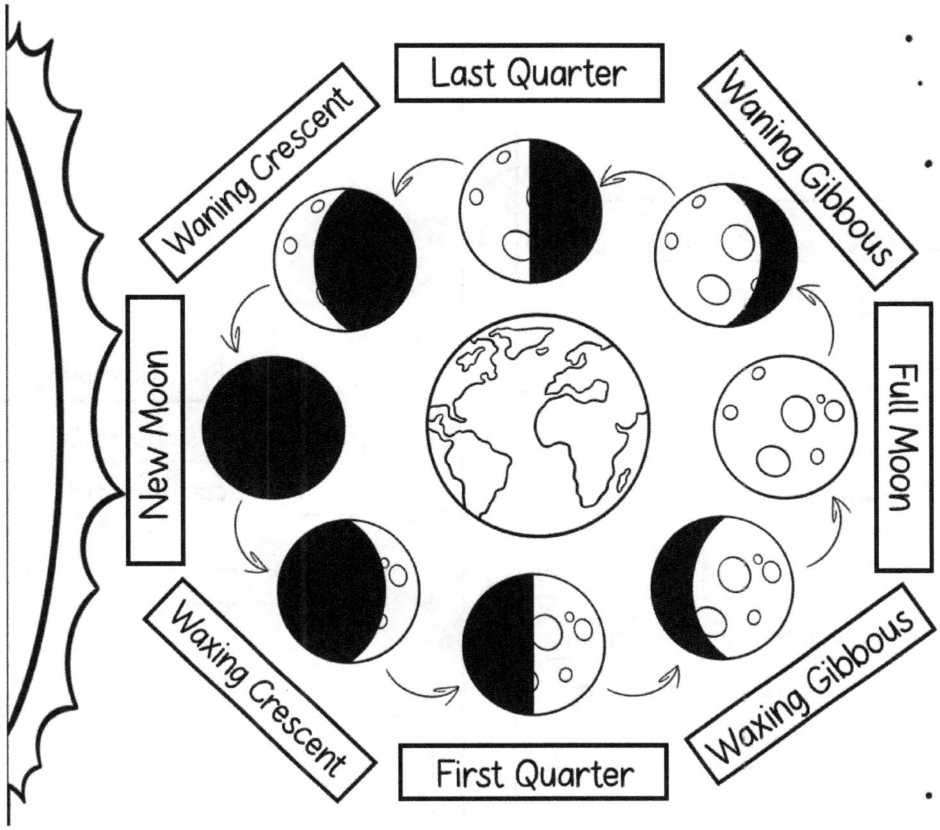

Page 289 – Day 3

Comets, Meteors, and Asteroids

C	E	C	A	D	U	E	N	E	C	O	M	A	T
A	S	E	P	S	O	P	E	I	T	I	M	S	G
M	H	A	B	D	T	U	M	A	S	I	I	N	B
O	O	W	D	C	M	E	T	E	O	R	S	L	E
U	W	L	M	A	C	S	R	O	N	E	R	G	H
F	E	A	S	R	P	O	Y	O	I	C	A	L	A
L	R	O	Y	S	R	T	M	R	I	I	T	N	V
A	I	S	P	Y	E	O	E	E	T	D	I	S	I
H	O	R	A	M	D	L	T	T	T	O	O	F	O
E	I	W	E	E	A	B	E	E	I	S	N	S	R
Y	A	S	E	R	T	I	O	O	R	O	F	I	T
D	S	B	T	Y	O	V	R	U	Z	Z	N	I	L
U	H	I	B	O	R	N	I	T	I	O	B	Q	N
S	R	A	V	F	R	Y	T	U	V	R	G	Z	L
T	H	O	J	E	R	Y	E	N	O	A	D	F	R
E	H	K	L	I	U	O	F	X	L	O	B	Y	I

History of Space Exploration

Page 316 – Day 4

OCTOBER 4, 1957
Event: Sputnik 1 launched
Fact: Sputnik 1 was the first man-made satellite to orbit Earth.

NOVEMBER 3, 1957
Event: Sputnik 2 launched
Fact: Laika was aboard Sputnik 2 and became the first animal to orbit Earth.

JANUARY 31, 1958
Event: Explorer 1 launched
Fact: Explorer 1 was the first American satellite, which discovered the Van Allen belts.

JULY 20, 1969
Event: Apollo 11 Moon landing
Fact: Neil Armstrong became the first human to touch foot on the Moon.

APRIL 11-17, 1970
Event: Apollo 13 mission
Fact: An oxygen tank explosion forced the mission to be aborted, but the crew returned safely.

APRIL 24, 1990
Event: Hubble Space Telescope launched
Fact: A powerful telescope that has captured stunning images of our universe.

Certificate of Achievement

This certificate is proudly awarded to:

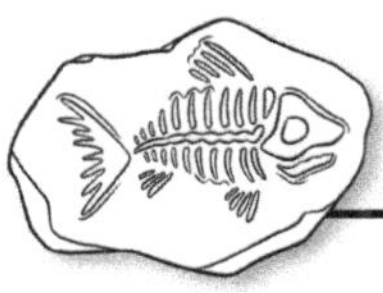

____

For successfully completing the Exploring Science Workbook.

Your hard work has paid off!

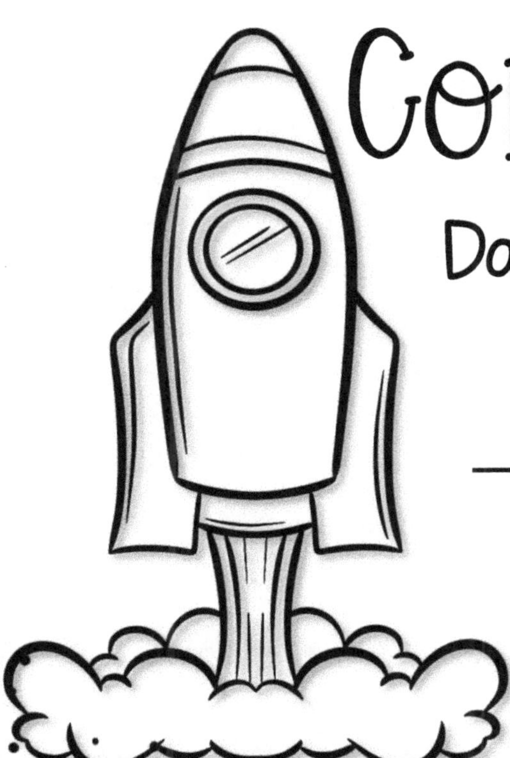

Congratulations!

Date: ______________________

____

Signature

Looking for more science resources to supplement your child's learning?

Our 'Science Throughout the Year' unit is designed to complement your child's daily science workbook plus additional topics. It provides over 1000 pages and slides of interactive activities, editable PowerPoint slides, lessons, posters, etc.
(ALL DIGITAL). Nothing will be mailed to you. You would receive a zip file with all of the editable PowerPoint slides, PDFs, and links to Google Slides.

Reinforce key concepts throughout the year. Simply scan the QR code below to access the unit and enhance your child's science education.

SCAN ME

What teachers are saying:

★★★★★

This resource helped me build confidence to teach science! I've always felt very intimidated and overwhelmed. These resources were so engaging! These made teaching science fun for me, which translated to engagement for my kids! I cannot say thank you enough!

- THE TRAVELING TEXAN TEACHER

★★★★★

I love this resource! The slides are beautiful and full of information. It is very well organized, the activities are engaging! The children loved all of the activities. I don't know if it had to do anything with this resource but, my students scored so much better on the FCAT than previous classes. Thank you!

- JOANN SO.

★★★★★

This was one of my best investments this year! My students loved the material and it helped make lesson planning much easier!

-SAMANTHA H.

Written by Fatema Ghazi
Illustrations by Mariam Ghazi

Contact the Author:
Fatema@thediscoveryapple.com
Contact the Illustrator:
Themagicalgallery@outlook.com

Thank You!

Thank you so much for choosing this workbook. As a Texas-certified teacher and curriculum designer, I created this resource to support meaningful, engaging learning for children. I truly hope it has been helpful for your family or classroom.

Your Review Makes a Difference

If you found this workbook helpful, I would truly appreciate it if you could take a moment to leave a quick review on Amazon. Even one sentence helps other parents and teachers decide if this resource is right for them and helps me continue creating high-quality learning materials.

Thank you for your support and for being part of this learning journey.

Warm regards,
Fatema Ghazi

www.ingramcontent.com/pod-product-compliance
Lightning Source LLC
LaVergne TN
LVHW062144200926
848484LV00007B/1854